Pedro Tersariol
André Rosa

Environmental diagnosis of the Córrego do Bugre watershed

Pedro Tersariol
André Rosa

Environmental diagnosis of the Córrego do Bugre watershed

A case study of a watershed belonging to the middle reaches of the Tietê/Sorocaba River

ScienciaScripts

Cover image: www.ingimage.com

This book is a translation from the original published under ISBN 978-3-330-75917-6.

Publisher:
Sciencia Scripts
is a trademark of
Dodo Books Indian Ocean Ltd. and OmniScriptum S.R.L publishing group

120 High Road, East Finchley, London, N2 9ED, United Kingdom
Str. Armeneasca 28/1, office 1, Chisinau MD-2012, Republic of Moldova, Europe
Printed at: see last page
ISBN: 978-620-8-30679-3

SUMMARY

DEDICATORY

To the family that encourages me, especially my parents Ivarne Tersariol and Ilde Santos Tersariol, my wife Narally Venturelli Gonfiantini Tersariol and son Luiz Henrique Gonfiantini Tersariol.

To every society that occupies a space in a sustainable way and understands it as an organic system, respecting other forms of life throughout the process of land use and occupation.

ACKNOWLEDGMENTS

First of all, I thank God, who has accompanied me throughout my personal and professional journey, enlightening me and giving me the strength to overcome all my challenges.

To my advisor, Prof. Dr. André Henrique Rosa, for his attention and dedication throughout these two years of the course.

To the laboratory technicians, especially Leticia Boschini Fraga Gonçalves and Suzan da Silva Lessa, for all their technical support and practical teachings on chemical analysis.

To all my teachers and friends in the Postgraduate Program in Environmental Sciences, especially Claudia Hitomi Watanabe for all her help and attention. I had the pleasure of meeting incredible people who have contributed a lot to my life.

To my family, who have always believed in me and in the possibilities and wonders that education provides for an individual and society in general.

To my work colleagues who always encouraged me during difficult times.

The city of Aluminio, which was the scene of this work.

"We have to realize (...) that there is no other planet to turn to, or to which we can export our problems. Instead, we have to learn (...) to live within our means." Jared Diamond

SUMMARY

Human growth since the Industrial Revolution and the post-war period has put drastic pressure on natural resources. In Aluminio - SP, the municipality's large urban concentration is densely packed into the sub-basin of the Bugre stream, greatly compromising the quality of its waters, especially in the urban perimeter. In this sense, environmental diagnosis is fundamental for monitoring and understanding the state and quality of the natural resources exploited and for helping to make decisions. Over the course of a seasonal year, this study carried out a diagnosis of water quality at three points along the Bugre stream, using parameters included in CETESB's IVA (Water Quality Index for the Protection of Aquatic Life and Aquatic Communities) and IQA (Water Quality Index). The first point is located at the source (an area with native vegetation) and the second and third are located in urban areas. The parameter data was statistically analyzed using Principal Component Analysis (PCA). According to the evaluations of the points, there were notable variations in water quality between the three points analyzed, which occurred due to their different geographical positions and also the influence of seasonality on the concentration of pollutants. The second and third points showed very high levels of various pollutants, especially total phosphorus and nitrogen, surfactants, fecal coliforms and chlorophyll-a, as well as some types of metal, such as copper and lead. The PCA indicated that the IET (Trophic State Index) was mainly related to phosphorus, the IQA (Water Quality Index) to DO and the IPMCA (Index of Minimum Variables for the Protection of Aquatic Life) to the presence of surfactants. The results show the need to restore its physico-chemical and microbiological characteristics in urban areas. Solutions such as sewage treatment are indispensable for this recovery. Environmental planning (supported by environmental zoning) in the context of decision-making by the public authorities is crucial if measures to restore this stream are to be implemented.

Keywords: diagnosis, indices, water quality.

1 INTRODUCTION

The growth in human population that occurred after the Second World War demanded greater consumption and, consequently, greater exploitation of the natural resources that are essential for guaranteeing and maintaining life on Earth, such as water resources.

The Bugre stream, which is an important source of water for the Sorocaba river basin, has been suffering for at least two decades from the pollution of its waters, resulting from intense population growth combined with a lack of consistent environmental planning. The anthropogenic impact is characterized by the discharge (clandestine or not) of untreated domestic sewage into this watercourse, mainly in the urban perimeter.

According to the director of Aluminio's Municipal Planning and Works Department, only industrial sewage is treated; domestic sewage is discharged directly into two watercourses: the Bugre stream and the Varjao stream (CRUZEIRO DO SUL, 2014).

Aquatic ecosystems end up, in one way or another, serving as temporary or final reservoirs for a wide variety and quantity of pollutants discharged into the air, soil or directly into bodies of water. In this way, pollution of the aquatic environment, caused directly or indirectly by man, through the introduction of inorganic or organic substances, produces deleterious effects such as: damage to living beings; danger to human health; negative effects on aquatic activities (fishing, leisure, among others) and damage to water quality with regard to use in agriculture, industry and other socio-economic activities (JUNIOR et al 2008). According to Refosco (1996), part of the effects of pollution are neutralized or stabilized by the receiving body, depending on the proportion of the mixture (dilution) and the natural stabilization potential of the water. However, anything that exceeds this capacity must be eliminated through appropriate treatment.

Julio et. al (2008) state that domestic sewage contains approximately 99.9% water and only 0.1% solids, which cause contamination or pollution of the water, generating the need to treat this sewage. Although sewage causes so many problems such as pollution, bacterial contamination, the appearance of diseases, among others, its treatment is non-existent in most Brazilian municipalities. The cost of installation and maintenance is the biggest obstacle to its viability, according to its administrators, even though the collection network is a frequent request from the community, because it removes sewage from their doorsteps (AISSE, 2000).

Today's technological components suggest a new concept of sanitation involving the tripod formed by man, nature and physical works, without one aspect predominating over the others, which is fundamental to achieving a level of social dignity. Thus, two aspects are fundamental to a new approach: environmental education and the sustainability of human development (JULIO et. al. 2008).

From this perspective, environmental monitoring and diagnosis are tools that help to monitor quality and propose more precise solutions for certain resources. With this, it is possible to raise awareness and mobilize the population around their environment, in order to guarantee the conservation of natural resources under different types of use and management.

The main objective of this work was to carry out environmental monitoring and diagnosis of the Bugre stream by means of the IVA (Water Quality Index for the Protection of Aquatic Life and Aquatic Communities) and IQA (Water Quality Index), with the aim of pinpointing the levels and parameters of the state of quality of this stream through the variables presented by these indices. The specific objectives were thematic mapping of land use and contour lines in the Bugre stream basin; water sampling at three points in the stream; measuring the flow of the stream at the third sampling point, near its mouth; the determination and elaboration of diagnostic quality indices (IET, IPMCA, IQA and IVA) related to aquatic life and public supply and the comparison of the results obtained with the limits of CONAMA Resolution 357/05 for class 3 watercourses.

1.1 Characterization of the study area.

The administrative formation of Aluminio began in 1942, when it became a police district. Later, in 1980, it became a district of the municipality of Mairinque. The municipality was created in 1991. Its origins and economic development are associated with the Sorocabana Railroad and Japanese immigration (FUNDAÇÂO FLORESTAL, 2012).

It borders the municipalities of Mairinque to the east and north; Ibiûna to the south; Votorantim to the southwest and Sorocaba to the west. The municipality of Aluminio has 16,839 inhabitants and a land area of 84 km^2 , which corresponds to a population density of 200.92 inhabitants per square kilometer (IBGE, 2010). Aluminio is located 75 km from the capital of São Paulo, on the Raposo Tavares Highway (SP - 270), and has the following geographical coordinates: latitude 23°32'6" S; longitude 47°15'40" W. It has an average altitude of 800 meters. According to the Koppen climate classification, Aluminio's climate type is Cwa, which covers the entire central part of the state and is characterized by a tropical high-altitude climate, with rainfall in summer and drought in winter, with the average temperature of the hottest month exceeding 22°C (CEPAGRI, 2013). The Bugre stream hydrographic basin is located practically in the entire urban perimeter of Aluminio - SP, which is part of UGRHI 10 - Mèdio Tietê e Sorocaba, in the southeast of the state of São Paulo, specifically making up the Alto and Médio Sorocaba sub-basins (IPT, 2006).

The municipality of Aluminio is part of the Atlantic Plateau. The Atlantic Plateau is characterized geomorphologically as a region of highlands, around 700 to 800 meters above sea level on average, corresponding to a geological shield where crystalline rocks of Precambrian to Cambrian-Ordovician age predominate, with the Unit cut by intrusive basic and alkaline rocks of Mesozoic to Tertiary age (IPT, 1981). The unit is covered by sedimentary rocks from the Sâo Paulo and Taubaté basins and bordered to the east by the Peripheral Depression Geomorphological Province made up of Paleozoic sedimentary rocks from the Paranà Basin. It should be noted that the transition between the two units (shield/basin) is often masked, and their boundaries are not readily apparent. The Atlantic Plateau has 13 Geomorphological Zones, namely the Atlantic Plateau, the Juqueriquerê Plateau, the Paraitinga Plateau, the Bocaina Plateau, the Middle Paraiba Valley, the Serra da Mantiqueira, the Upper Rio Grande Plateau, the Sao Roque Mountains, the Jundiai Plateau, the Lindóia Mountains, the Ibiùna Plateau, the Guapira Plateau and the Upper Rio Turvo Plateau. Aluminio is geologically located in the Pirajibu sub-basin, making up the Sorocaba river basin (figure 1) (IPT, 1981).

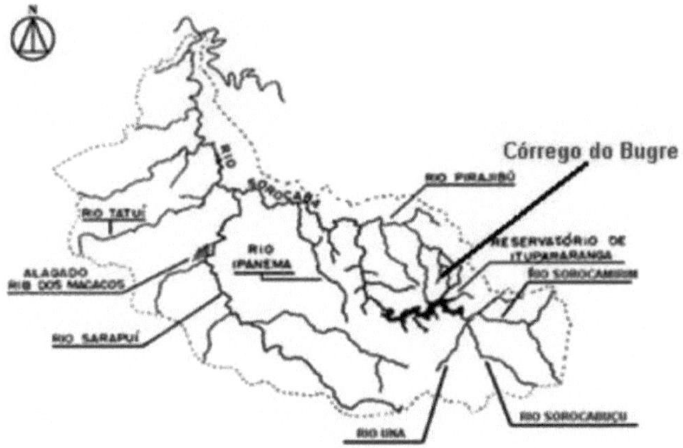

Figure 1 - Location of C. do Bugre in the Sorocaba River sub-basin.

Source: SMITH et al. (2005) adapted.

The Pirajibu sub-basin is located near the eastern edge of the Paranà Sedimentary Basin, occupying mostly crystalline rocks belonging to the southeastern folding region, the Ribeira Belt, the Sâo Roque Block and hydrologically to the Middle Tietê/Sorocaba Basin, where the large crystalline aquifer system occurs. The lithological units that make up the Sâo Roque Block in the sub-basin mainly comprise the Sâo Roque and Serra do Itaberaba Groups, formed predominantly by metasediments, and the Sorocaba and Sâo Francisco Granitoid Massifs, partially covered by sedimentary rocks of the Itararè Subgroup (SILVA ET. al., 2003).

2. LITERATURE REVIEW

The belated concern with the environment and consequently with the state of natural resources is still a major challenge for society and today's leaders. On the world stage, it began systematically in the 1970s, with the creation of the EPA (United States, 1970), the Stockholm Conferences (Sweden, 1972) and Tbilisi (Georgia, 1976). In Brazil, the environmental issue began to be better discussed in the 1960s, with the Forest Code Law (4771/65), but it was only in the 1980s, with the National Environmental Policy (6938/81) and the 1990s, with the United Nations Conference on Environment and Development (RIO-92) and the creation of the National Water Resources Policy (9433/97) that this issue became more intensely manifested, respectively.

Still on the national scene, Article 2 of the National Environmental Policy (6938/81) (MMA, 2013), guarantees the preservation, improvement and recovery of environmental quality conducive to life, aiming to ensure, in the country, conditions for socio-economic development, the interests of national security and the protection of the dignity of human life.

With regard to water resources, Brazil currently has a large legal framework, guaranteed by Law 9433/97 (MMA, 2013). However, these policies are not carried out (apart from exceptions) in harmony between municipalities, states and the union. Furthermore, in the Brazilian context, the majority of water sources located in urban areas suffer from the neglect of public administration and population growth. There are many clean-up projects downstream of the main dumping points, in the main rivers, but there is no obvious concern about solving the root of the problem, which is, among other things, the industrial or domestic dumping points, clandestine or not, upstream of the main rivers.

The diagnosis of environmental problems is indispensable for planning and more effective decision-making. Environmental planning of the territory (or of a river basin) becomes both a basic and complementary element for economic and social development and for optimizing the use, management and management plan of any territorial unit (SOARES, 2004).

According to Santos (2004), the inventory and diagnosis of environmental planning represent the way to understand the potential and weaknesses of the study area, the historical evolution of occupation and the pressures of man on natural systems. They also shed light on the successes and conflicts of land use and past, present and future impacts. These assessments consider temporal, spatial and scalar variations, in a back and forth process, in various combinations. Portraits of the area are formed which, when compared, added together and interpolated, highlight the main characteristics and provide indications of the dynamics of the region. Depending on the methodology used, territorial or landscape units, zones or scenarios can be used. In reality, the aim is to compartmentalize space into planning and management units. Each unit should correspond to a specific set of alternatives and actions.

There is often confusion between the concepts of planning, management and environmental zoning, as if they were synonymous. Planning takes into account the taking of measures; management is the control and monitoring of something. Finally, zoning is the establishment of zones, i.e. within the environmental context, zoning is territorial planning in homogeneous zones with different potentials or weaknesses. This means that environmental planning depends exclusively on management and zoning (ZACHARIAS, 2006).

Environmental zoning, made up of a set of methodologies (including thematic mapping of land use and occupation) is indispensable, for example, in assessing potential risk areas, representing both continuous and dynamic characteristics of the study area. Environmental management, which can be carried out seasonally by monitoring the physical, chemical and biological analysis of the Bugre stream, provides a temporary assessment of the quality of its waters. Both are covered by the City Statute, which in turn is governed by the Municipal Master Plan (mandatory only in cities with 20,000 inhabitants or more).

2.1 Field survey

Sampling is a very important stage in assessing the area in question, and so sampling must be carried out in a careful and technical manner in order to avoid contaminating samples and thus be representative of the body of water being sampled (ANA, 2011). According to the National Water Agency - ANA (2011), the collection of water samples must take certain precautions, such as checking the cleanliness of the materials used for depositing it and making sure that they are in harmony with the local water, not contaminating the samples with impurities, among other recommendations, and depends on three factors:

- The matrix to be sampled, i.e. whether the body of water is shallow or not, whether it is treated, among others;
- The type of sampling, i.e. whether it is a simple, composite or integrated sample;
- The tests to be carried out later, i.e. physical-chemical and microbiological tests.

In order to guarantee the homogeneity and representativeness of the proposed sampling site, the actions to be taken must be carefully planned. There are currently hundreds of variables or determinants that can be used to characterize a body of water, involving physical, chemical, microbiological, biological, toxicological and radiological parameters. These parameters must be defined with adequate knowledge of their meaning, scope, limitations, reliability, references for comparison and the costs of obtaining them. The combinations of these variables do not allow standard plans to be formulated. Each case must be studied individually, and the most commonly used parameters and criteria include those established in current legislation (CETESB, 2011).

The variables mentioned change mainly over space and time. According to CETESB (2011), the intensity of these variations can be reduced, for example, as the sampling point moves away from the point of release. Therefore, in order to establish the time and frequency of sampling, the temporal variability of each parameter must be known for each sampling site. Based on the profile of this variability, it is possible to establish the sampling program and the number of samples that should be taken. The greater the number of samples investigated, the better the knowledge of variability and, consequently, the estimate of environmental impact.

2.2 VAT, IPMCA, IET and IQA.

2.2.1 VAT variables

The purpose of VAT is to assess water quality for the protection of fauna and flora in general, and it is therefore different from an index for assessing water for human consumption and primary contact recreation. IVA takes into account the presence and concentration of toxic chemical contaminants, their effect on aquatic organisms (toxicity) and two of the variables considered essential for biota (pH and dissolved oxygen). These variables are grouped into the IPMCA - Index of Minimum Variables for the Preservation of Aquatic Life and the IET - Carlson's Index of Trophic State modified by Toledo (1990). In this way, the IVA through the weighting degrees (table 1) provides information not only on the quality of the water in ecotoxicological terms, but also on its degree of trophy (CETESB, 2007).

Table 1 - VAT weightings

Category	Weighting
Great	VAT ≤ 2.5
Good	2.6≤ VAT ≤ 3.3
Regular	3,4 ≤ IVA ≤ 4,5
Bad	4.6≤ VAT ≤ 6.7
Very bad	VAT ≥ 6.8

The IPMCA is made up of two groups of variables (table 1)

1) **The group of essential variables** (dissolved oxygen, pH and toxicity). For each variable included in the IPMCA, three different quality levels are established, with numerical

weightings from 1 to 3 and which correspond to water quality standards established by CONAMA Resolution 357/05, and standards recommended by American (USEPA, 1991) and French legislation (Code Permanent: Environnement et Nuisances, 1986), which establish maximum permissible limits for chemical substances in water, with the aim of avoiding the effects of chronic and acute toxicity on aquatic biota (CETESB, 2007).

2) Group of toxic substances (copper, zinc, lead, chromium, mercury, nickel, cadmium, surfactants and phenols). This group includes the variables currently assessed by the São Paulo State Inland Water Quality Monitoring Network and which identify the level of contamination by substances potentially harmful to aquatic communities.

These levels reflect the following water quality conditions (table 2). Level A: Waters with desirable characteristics to maintain the survival and reproduction of aquatic organisms. Meets the quality standards of CONAMA Resolution 357/2005 for class I and II waters - (weighting 1). The exceptions are Dissolved Oxygen (DO) for class I whose value is =6.0 mg/L O2 and Total Phenols. Level B: Waters with desirable characteristics for the survival of aquatic organisms, but reproduction may be affected in the long term (weighting 2). Level C: Waters with characteristics that may compromise the survival of aquatic organisms (weighting 3).

Groups	Variables	Levels	Range of variation	Weighting
Essential variables (PE)	OD (mg/L)	A	≥ 5	1
		B	3 a 5	2
		C	< 3	3
	pH	A	6 a 9	1
		B	5 to < 6 and > 9 to 9.5	2
		C	< 5 e > 9,5	3
	Toxicity	A	Non-toxic	1
		B	Chronic effect	2
		C	Acute effect	3
Toxic substances (TS)	Cadmium (mg/L)	A	≤ 0,001	1
		B	> 0,001 a 0,005	2
		C	> 0,005	3
	Chromium (mg/L)	A	≤ 0,05	1
		B	> 0,05 a 1	2
		C	> 1	3
	Copper (mg/L)	A	≤ 0,02	1
		B	> 0,02 a 0,05	2
		C	> 0,05	3
	Lead (mg/L)	A	≤ 0,03	1
		B	> 0,03 a 0,08	2
		C	> 0,08	3
	Mercury (mg/L)	A	≤ 0,0002	1
		B	> 0,002 a 0,001	2
		C	> 0,001	3
	Nickel (mg/L)	A	≤ 0,025	1
		B	> 0,025 a 0,160	2
		C	> 0,160	3

	Phenols (mg/L)	A	≤ 0,001	1
		B	> 0,001 a 0,050	2
		C	> 0,050	3
	Surfactants (mg/L)	A	≤ 0,5	1
		B	> 0,5 a 1	2
		C	> 1	3
	Zinc (mg/L)	A	≤ 0,18	1
		B	> 0,18 a 1	2
		C	> 1	3

Table 1 - IPMCA component variables

Table 2 - IPMCA weightings

Category	Weighting
Good	1
Regular	2
Bad	3 e 4
Bad	≥ 6

Toxicity is measured based on an assessment of the essential variables and the group of toxic substances.

2.2.3 EIT variables.

Eutrophication is the enrichment of a body of water, the artificial form of which is harmful to the system and occurs over a short period of time. This process tends to compromise the quality of the water and the ecosystem by leading to the following consequences: reduction in dissolved oxygen concentrations, causing the death of many organisms, including fish; proliferation of phytoplankton biomass; increase in potentially toxic cyanobacteria populations; proliferation of aquatic macrophytes, which can clog pipes; and other problems, such as siltation and foul odors (TUNDISI; MATSUMURA-TUNDISI, 2008).

The purpose of the Trophic State Index (TSI) is to classify bodies of water into different degrees of trophy (table 3), i.e. it assesses water quality in terms of nutrient enrichment and its effect on excessive algae growth or increased infestation of aquatic macrophytes. In this index, the results corresponding to total phosphorus, IET (Pt), should be understood as a measure of the potential for eutrophication, since this nutrient acts as the causative agent of the process. The assessment corresponding to chlorophyll-a, EIT (CL), in turn, should be considered as a measure of the water body's response to the causative agent, adequately indicating the level of algal growth taking place in its waters. Thus, the average index satisfactorily encompasses the cause and effect of the process (CETESB, 2007).

Table 3 - Trophy level classification according to Carlson's calculation (1977) modified by Lamparelli (2004)

IET CLASSIFICATION	
Degree of trophy	Min - Max
Ultraoligotrophic	0 - 47

Oligotrophic	47 - 52
Mesotrophic	52 - 59
Eutrophic	59 - 63
Supereutrophic	63 - 67
Hypereutrophic	> 67

2.2.3 IQA variables

The IQA is intended to serve the public supply. It can be calculated using nine parameters (table 4), where water quality variation curves are established (where each parameter has a different weight from the other) according to the state or condition of each one (CETESB, 2007).

Table 4 - Weights of the parameters analyzed by IQA

IQA parameters	Weights (w)
OD (mg/L)	0,17
pH	0,12
BOD (mg/L)	0,1
Thermotolerant coliforms (MPN/100 ml)	0,15
Temperature (°C)	0,1
Turbidity (UNT)	0,08
Total residue (mg/L)	0,08
Total phosphorus (mg/L)	0,1
Total nitrogen (mg/L)	0,1

The IQA is a number between 0 and 100, and its weightings are shown in Table 5 (CETESB, 2007).

Table 5 - IQA weighting grades

IQA classification	
Category	Weighting
Great	79 <IQA ≤ 100
Good	51 < IQA ≤ 79
Regular	36 < IQA ≤ 51
Bad	19 < IQA ≤ 36
Bad	IQA ≤ 19

2.3 Using ICP OES to determine metals

The inductively coupled plasma atomic emission spectroscopy (ICP-OES) technique is widely used when low concentrations of metals and metalloids need to be determined, due to its good sensitivity, precise and accurate measurements and low detection limits (PETRY, 2005).

The ICP OES spectroscopic technique used to read metals in this work is based on measuring the emission of electromagnetic radiation from atoms and ions generated by an argon (Ar) plasma. The plasma is an ionized gas, usually made up of argon, generated inside a compartment of the equipment called the torch compartment (PETRY, 2005).

Air plasma has important fundamental characteristics, including gas temperature (Tg), electron temperature (Te) and electronic density (ne). The plasma has a high gas temperature (4500 to 8000 K) and electron temperature (8000 to 10000 K) and, with a long residence time for the sample aerosol inside the plasma (2 to 3 ms), it vaporizes completely and the elements present are atomized/ionized. As a result, chemical interferences in the plasma are reduced and, due to the high electronic density in Air ICP, ionization interferences are small compared to flame spectrometry, where the electronic density is lower (PETRY, 2005).

Since the 1970s, when the first ICP OES equipment was installed in Brazil, its optical components and detection systems have been improved in order to obtain more accurate and precise results. A considerable example is the use of the axial view of the plasma (the emitted radiation is focused by the optical system along the central channel of the plasma), which provides better LDs than the radial view (only a small part of the radiation is focused), by about an order of magnitude, although the matrix effects are more pronounced. In addition, some instruments have both radial and axial views, improving the versatility of the technique. Matrix effects (sample constituents) can be minimized by using a more robust plasma, which is obtained by using high radio frequency (RF) power and low nebulizer gas flow, leaving the plasma in conditions where interference is less severe (PETRY, 2005).

2.4 Geoprocessing and environmental diagnosis

In the modern perspective of land management, any action to plan, organize or monitor space must include an analysis of the different components of the environment, including the physical-biotic environment, human occupation and their interrelationship. The concept of sustainable development, enshrined in Rio-92, establishes that land use actions must be preceded by a comprehensive analysis of their impacts on the environment in the short, medium and long term. In this way, we can point to at least four major dimensions of the problems linked to environmental studies that should be carried out together, where the impact of the use of Geographic Information Systems technology is great: Thematic Mapping, Environmental Diagnosis, Environmental Impact Assessment, and Territorial Planning (CAMARA & MEDEIROS, 2001).

The term Geoprocessing denotes the discipline of knowledge that uses mathematical and computational techniques to process geographic information. This technology is increasingly influencing the areas of Cartography, Natural Resource Analysis, Transportation, Communications, Energy and Urban and Regional Planning (CÂMARA & MEDEIROS, 2001).

Geoprocessing computer tools, known as Geographic Information Systems (GIS), make it possible to carry out complex analyses by integrating data from various sources and creating geo-referenced databases. They also make it possible to automate the production of cartographic documents. In a country as continental in size as Brazil, with a great lack of adequate information for decision-making on urban, rural and environmental problems, Geoprocessing has enormous potential, especially if it is based on relatively low-cost technologies, in which knowledge is acquired locally (CAMARA & MEDEIROS, 2001).

2.5 Flow as a hydrological indicator

Flow is the volume of water that passes through a given section of river per unit of time, which is determined by the variables of depth, width and flow velocity, and is commonly expressed in the international system (SI) of measurements in m^3/s. In a stream, from one bank to the other and from the surface to the bed, the flow is not homogeneous, so this also implies a variation in the discharge (flow), which varies in the vertical and transversal sections of the river. This is due to the morphology of the river, in which the friction of the water on the banks and bed causes a slowing effect on the speed, as well as the friction effect of the surface water with the atmosphere (CARVALHO, 2008).

The methods used to determine flow can be indirect or automatic, ranging from a simple object thrown into the water to estimate the speed it travels over a given distance, to more precise methods such as windlasses, acoustic dopplers (ADCP - *Automatic Doppler Current Profiler*). Among these, the use of the hydrometric windlass is the most widespread, due to its ease and cost-effectiveness (CARVALHO, 2008).

2.6 Water parameters analyzed

Conductivity: conductivity is a parameter that reflects the capacity of water to conduct electric current and is highly related to the amount of total dissolved solids and dissolved ions in the water, thus reflecting the state of inorganic pollution in the aquatic environment (JONNALAGADDA & MHERE, 2001).

Dissolved oxygen (DO): one of the most important constituents for aquatic species is DO. The concentration of DO is directly related to the forms of organisms that can survive in a body of water (MANAHAN, 2005). For the existence of higher forms of aerobic life, a minimum dissolved oxygen concentration of 2 mg L^{-1} is calculated to be necessary, with more demanding species requiring a minimum of 4 mg L^{-1} (BALLS, et. al. 1996). High concentrations of dissolved oxygen indicate the presence of photosynthetic plants and low values indicate the presence of organic matter or other reducing agents. In this way, dissolved oxygen becomes one of the main parameters for controlling water pollution levels. Its determination is fundamental for maintaining and verifying the aerobic conditions of a body of water that receives polluting material (ROSA, et. al. 2009).

Hydrogen potential (pH): pH is a very important characteristic of water, as its value has a great influence on many chemical reactions that occur in the environment, directly and indirectly affecting the physiology of species (BRAGA, et. al. 2005). The pH value can be significantly altered when various substances are dumped into the aquatic environment as a result of human activity, such as acidic deposits from atmospheric pollution (BRAGA, et. al. 2005).

Temperature: Natural bodies of water show temperature variations that are part of their normal climatic regime, with seasonal and diurnal variations, as well as vertical stratification. Factors such as latitude, altitude, seasonality, time of day, flow rate and depth have a significant influence on temperature (ROSA, et. al. 2009). Temperature plays an important controlling role in the aquatic environment, influencing the behavior of a series of physico-chemical parameters. The solubility of gases and the kinetics of chemical reactions are altered by temperature, meaning that the interaction of pollutants with the aquatic ecosystem is greatly influenced by its variation (BRAGA *et al.,* 2005). In addition, many aquatic life forms have limited thermal tolerance and optimum temperatures for reproduction and growth. Large rises in the temperature of a body of water are usually caused by anthropogenic activities.

Fecal coliforms: Water is the biggest source of enteric diseases, causing numerous illnesses every year, such as cholera and diarrhea. This contamination generally occurs through pathogenic organisms that are discharged into water bodies via untreated urban effluents and whose water is consumed by the population living around the watershed (EGWARI & ABOABA, 2002).

Chlorophyll-a: chlorophyll-a is concentrated when the concentration of eutrophying nutrients in the water increases. Phytoplankton biomass grows in accordance with the existence of these nutrients and directly with the increase in sunlight, which is the limiting factor for algal growth. This parameter varies according to the suspended sediments and the concentration of phytoplankton. The higher the turbidity value, the more it limits the entry of light, thus limiting the primary productivity of algae and, consequently, the concentration of chlorophyll-a (CULLEN, 1982).

Biochemical oxygen demand (BOD $_{,520}$): the biodegradation of organic matter naturally present in aquatic ecosystems occurs due to the presence of microorganisms in the water, which consume oxygen to obtain energy. In this context, the quantification of the biochemical oxygen demand for 5 days at 20°C (BOD5$_{.20}$) can be understood as an indicator of the amount of oxygen that is consumed by the microorganisms in the body of water, whose purpose is to degrade the organic matter present in domestic and industrial effluents and from leaching in the watershed (ROSA et al., 2004).

Total solids: the presence of solids can cause harm to fish and aquatic life. This occurs as solids settle on the river bed, destroying organisms that provide food, or damaging fish spawning beds. Solids can trap bacteria and organic waste at the bottom of rivers, promoting anaerobic decomposition. Suspended particulate matter comes from atmospheric dust removed by rain, contact with soil, plant fibers, decaying vegetation, aquatic animal waste, algae, plankton, sediment resuspension, etc. After a storm there is a high concentration of inorganic sand. In summer there is little sand, but lots of algae and other aquatic organisms. Suspended particles make the water look bad, increasing turbidity and color, and reducing light penetration (CETESB, 2008).

Turbidity: turbidity is a parameter that indicates the degree of attenuation that a beam of light suffers when it tries to pass through a sheet of water, which indicates the presence of suspended solids, often caused by erosion of the banks of water bodies or contamination by urban effluents. Thus, these parameters are considered indirect indicators of the degree of purity of the water in a given body of water (SOUZA et al., 2007).

Total phosphorus (Pt): its presence in water bodies is most often the result of direct discharges, such as domestic or industrial effluents, not excluding the possibility of its presence due to soil washing in agricultural areas (ANDREWS et al., 2004). In domestic effluents, superphosphate detergents used on a large scale are the main source, in addition to the fecal matter itself, which is rich in protein (CETESB, 2008).

Total nitrogen (Nt): there are several sources that contribute to the increase in the concentration of nitrogen in natural waters. Sanitary sewage is generally the main source, discharging organic nitrogen and ammoniacal nitrogen due to the presence of proteins and the hydrolysis undergone by urea, respectively. Some industrial effluents also contribute to nitrogen discharges into water (ROSA, et. al. 2009).

Surfactants: among the numerous sources of these pollutants in the environment, we can highlight the emissions of urban effluents containing high concentrations of synthetic detergents, which have been used more intensively in recent years, since they are more soluble than ordinary soap (BIGARDI, 2003).

Phenols: phenols are long-chain organic compounds that are difficult to degrade. They are one of the main residues from oil refineries, plastic industries, paints and pesticides, making them one of the main and most problematic chemical pollutants found in aquatic ecosystems (SILVA & ASSIS, 2004).

Cadmium (Cd): the presence of cadmium in bodies of water is mainly due to industrial discharges, in particular the metal surface treatment industry, gas emissions from the burning of fossil fuels and the deposition of suspended materials in the atmosphere. It can also come from leaching from agricultural areas treated with pesticides containing this toxic metal (YANG & SANDO-WILHELMY, 1998). As a toxic metal, it biomagnifies in trophic chains, concentrating in the kidneys, liver, pancreas and thyroid. For humans, this metal can cause kidney dysfunction, hypertension, arteriosclerosis, among other pathologies (CACCIA et al., 2003).

Lead (Pb): Lead is a cumulative metal, like most toxic metals, and can lead to chronic poisoning called saturnism, which acts on the central nervous system. Contamination of the body by lead can cause muscle deficiency, gastrointestinal inflammation, vomiting and diarrhea in addition to saturnism (USEPA, 2009a).

Copper (Cu): because it is an important algaecide, pesticide, fungicide, among other harmful actions on organisms, copper is widely used in effluent treatment plants and in agriculture in general. Because of this intensive use, it can contaminate the water table through rainwater runoff and consequent leaching from agricultural land (ROSA, et. al, 2009).

Chromium (Cr): its concentrations are very low, generally less than 1 µg L^{-1} . It also comes from the treatment of metal surfaces. In its trivalent form it is essential for the proper functioning of the body, but in its hexavalent form it is extremely toxic (SHUKLA et al., 2008).

Mercury (Hg): water contamination by mercury can be caused by various routes, mainly direct emission into water bodies or even atmospheric deposition in the drainage basin. The consumption of fish contaminated with methyl mercury is one of the most important forms of exposure to mercury, which has consequences for the cardiovascular and nervous systems (QURESHI et. al., 2009). Bivalent mercury can be strongly absorbed by sediments and can be converted into elemental or methyl mercury, which is toxic and highly bioaccumulative (ULLRICH et al, 2001).

Nickel (Ni): Also used in metal surface treatment industries, some studies show that the presence of Ni can lead to carcinogenesis. Like zinc and copper, nickel, if present in aquatic environments, can precipitate in the gills of fish, killing them by asphyxiation, as well as being toxic when present in low pH solutions (ROSA, et. al. 2009).

Zinc (Zn): its access to aquatic ecosystems can be through water-rock-soil interaction, or through industrial processes that release it as effluent. The main industries that pollute aquatic ecosystems with this metal are steel mills, as well as the generation of domestic sewage, which has high concentrations of zinc. When present in high levels in bodies of water, it gives them a milky appearance, thus significantly increasing turbidity values (CETESB, 2008a).

3. OBJECTIVES

The main objective of this work was to carry out environmental monitoring and diagnosis of the Bugre stream by means of the IVA (Water Quality Index for the Protection of Aquatic Life and Aquatic Communities) and IQA (Water Quality Index), with the aim of pinpointing the levels and parameters of the state of quality of this stream through the variables presented by these indices.

3.1 Specific objectives

The specific objectives were:

- thematic mapping of land use and contour lines in the Bugre stream basin;

- water was collected at three points in the stream,

- measuring the flow of the stream in question at the 3rd collection point, close to its mouth,
- determining and drawing up diagnostic quality indices (IET, IPMCA, IQA and IVA) relating to aquatic life and public supply.

- comparison of the results obtained with the limits of CONAMA Resolution 357/05 for class 3 watercourses.

In this context, the objectives and proposals of this work fall within the scope of Art. 2 of Law 6938/81 and Art. 1 of Law 9433/97, using the hydrographic basin as the unit of study. With a view to better environmental planning of the Bugre stream basin, monitoring and diagnosing its waters are essential for understanding the current state of the quality of this resource.

4. MATERIAL AND METHODS

The following flowchart (figure 2) provides a more complex understanding of the main procedures carried out in this work

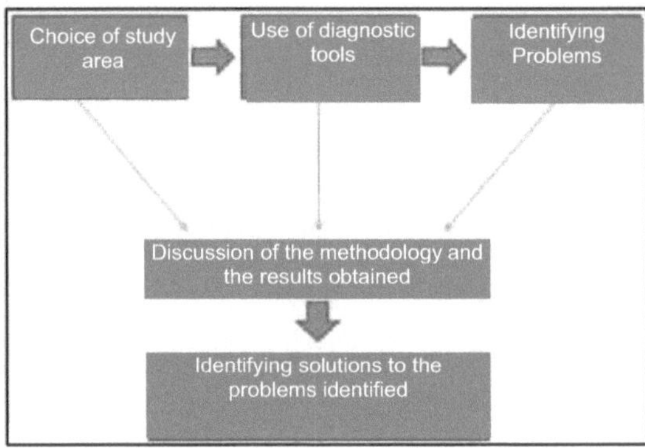

Figure 2 - Flowchart of the research stages.

4.1 Sample points and map generation.

In order to more accurately verify the effects of anthropogenic activities in the Córrego do Bugre basin, the assessment of water quality parameters at three points was carried out by means of IVA and IQA, the 1st point being upstream of the stream - main source (Bairro Itararé) - 23°34'31"S / 47°15'24.9"W; the 2nd point: near Aluminio City Hall (Centro) - 23°31'52.7"S / 47°15'11.5"W; and the 3rd point: near its downstream, 200 meters from its confluence with the Varjâo Stream and 10 meters from the Raposo Tavares Highway (Vila Pedàgio) - 23°31'06.6"S / 47°15'18.9"W (figure 3).

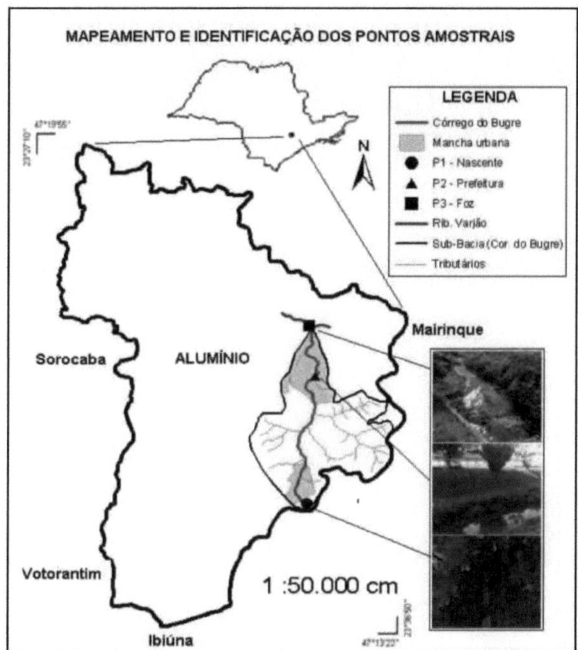

Figure 3 - Location of the sampling points.

The study area covers around 22 km^2 . The land use and contour lines with the average altitudes of the Bugre stream basin were mapped based on the IBGE topographic map of the municipality of Aluminio (2010) on a scale of 1:50,000. The map was digitized using Arcgis 10.0 software, using the ArcMap and ArcCatalog applications. The attributes of the figure were made in ArcCatalog with the datum SAD 69 - Córrego Alegre 23°S, and the attributes were digitized in ArcMap.

4.2 Determining the flow of the study area

Based on rainfall and with the aim of researching the variations in the drought and flood regime of the Bugre stream basin, flow tests were carried out with the Global Water FP211 windlass (figure 4) in the winter and summer months. The depth of each point was measured using the windlass's graduated rod, and the widths were obtained using a tape measure. According to Gomes and Santos (2003), water resources less than one meter deep require only one flow measurement in relation to their depth.

As velocity and depth (and therefore flow) vary between sampling points, it is necessary to measure them individually at each point, so that it is possible to obtain an average of the measurements taken (CARVALHO, 2008).

The windlass was inserted at 40% of the depth of each point, following the method in the Global Water Flow Probe manual for water resources up to 1 meter deep. In mathematical terms, the flow rate can be calculated using the formula:

$$Q = (w \cdot h) \cdot V \quad (1)$$

Where:

Q = flow (m^3/s)

V = water flow velocity (m/s)

h = average depth in the channel cross-section (m)

w = channel width (m)

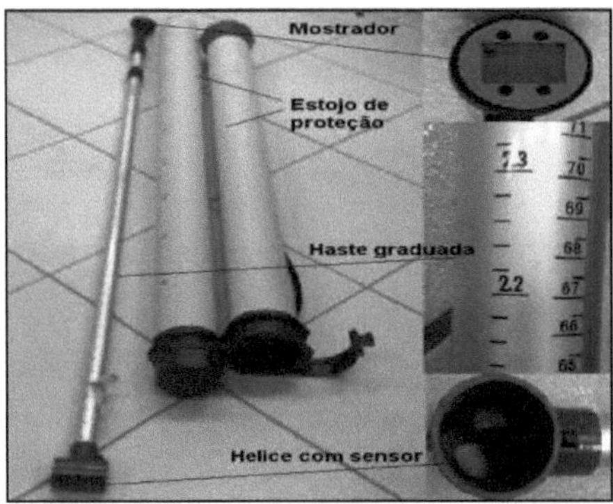

Figure 4 - Global Water FP 211 windlass

4.3 Equipment, reagents and methods used.

The samples were collected and preserved according to the criteria of the National Guide for the Collection and Preservation of Samples (ANA, 2011), respecting the criteria for each parameter analyzed. The samples were collected and preserved according to the criteria of the ANA National Guide to Sample Collection and Preservation (2011). Chlorophyll-a (CL) was determined according to the method of Wetzel and Likens (1991). Quantification was carried out using a HACH 3OR3900 spectrophotometer at wavelengths of 665 and 750 nm. Electrical conductivity was analyzed using a WTW LF 330 conductometer with a TetraCon® 325 conductivity cell. Thermotolerant coliforms (TC) were determined using the methodology described in "Standard Methods for the Examinations of Water and Wastewater", 21ª ed., 9222 B (APHA, 2005). The biochemical oxygen demand (BOD) was obtained using the methodology described in "Standard Methods for the Examinations of

Water and Wastewater" (APHA, 2005). To obtain total nitrogen (Nt), the samples were acidified with H2SO4 for preservation and digested with total nitrogen reagents from the HACH kit; read using a HACH DR2800 spectrophotometer (method 10071); total phosphorus (Pt) was determined by acidifying the samples with H2SO4 for preservation and digesting with persulfate; read using a HACH DR2800 spectrophotometer (method 8190). Dissolved oxygen (DO) and temperature (T) were determined using the WTW Oxi 315i oximeter with CellOx 325 sensor. A HACH 2000 spectrophotometer was used for turbidity readings. For surfactants, chloroform and reagents A and B from the HACH kit for anionic surfactants (substances reactive with methylene blue) were used; readings were taken using a HACH DR2800 spectrophotometer (LCK 332 method). To determine phenols, chloroform was used to extract the phenol compounds and reagent 1 and 2 for phenols, read using a HACH DR2800 spectrophotometer (method 8047). In order to evaluate total and dissolved metals (cadmium - Cd, copper - Cu, chromium - Cr, lead - Pb, mercury - Hg, nickel - Ni and zinc - Zn) in a sample, digestion must first be carried out, which consists of making the metal ions available in the solution, thus breaking their bonds with organic matter, for example. The sample was digested to evaluate total and dissolved metals. Sample readings were taken on an Agilent 700 series ICP-OES. Finally, total solids (TS) were determined using a muffle furnace, a drying oven at 105° C, a desiccator, an analytical balance, porcelain crucibles (50 mL) and a graduated pipette. Obtained using the equation:

$\mathbf{ST = (\mathit{b} - \mathit{a})/\mathit{V}\ g/mL}$ (convert to mg/L);$^{-1}$

2)

Where:

a = mass of the crucible before the sample was introduced, b = mass of the crucible with the dry residue after heating (105°C) and cooling the sample in the desiccator.

According to the methods mentioned above, the detection limits for total phosphorus, total nitrogen, surfactants and phenols are 0.04 - 3.5 (mg/L), 0.5 - 25 (mg/L), 0.05 - 2 (mg/L) and 0.002 - 0.2 (mg/L) respectively.

The following equipment, solutions, reagents (Table 2) and methods were used to obtain the analyzed parameters.

Chlorhydric Acid (F. Maia) 37% P.A
Nitric Acid (F. Maia) 65 % P.A
Sulfuric Acid (F. Maia) 98 % P.A
Analytical balance (accuracy 0.0001 g) SHIMADZU Mod. AW 220
TECNAL vacuum pump Mod. TE-058
FORT LINE gas extraction hood Mod. BR-CA 100
Heating plate, TECNAL Mod. TE-018

WTW LF 330 conductivity meter with TetraCon® 325 conductivity cell
HACH DR2000 spectrophotometer for turbidity
HACH DR2800 spectrophotometer
3OR3900 spectrophotometer
0.45µm Milipore® filter
American Lab microbiological incubator, AL 100
Test'n Tube® kit for total phosphorus determination (HACH)
Test'n Tube® kit for total nitrogen determination (HACH)
Gilson automatic micro pipettes, various volumes (fixed and variable)
WTW Oxi 315i oximeter with CellOx 325 sensor
WTW pH 315i pH meter
Reagent for detergents (surfactants) (HACH)
Reagent for Phenol (HACH)
Reagent for Phenol 2 (HACH)
CONTINENTAL Refrigerator Mod. 460
Deionizer system for water purification, MILIPORE Mod. Milli-Q Plus
Alkaline NH4 solution (pH=8.0) of 90% acetone
Glassware common to an Analytical Chemistry laboratory

Table 2 - Equipment and reagents used.

4.3.1 Preparation of total and dissolved metal samples.

In order to evaluate metals in a sample, digestion was first carried out, which consists of making the metal ions available to the solution, thus breaking the bonds with organic matter, for example. Digestion of the sample was carried out to evaluate total metals and dissolved metals. The samples were analyzed on an Agilent 700 series ICP-OES (figure 5).

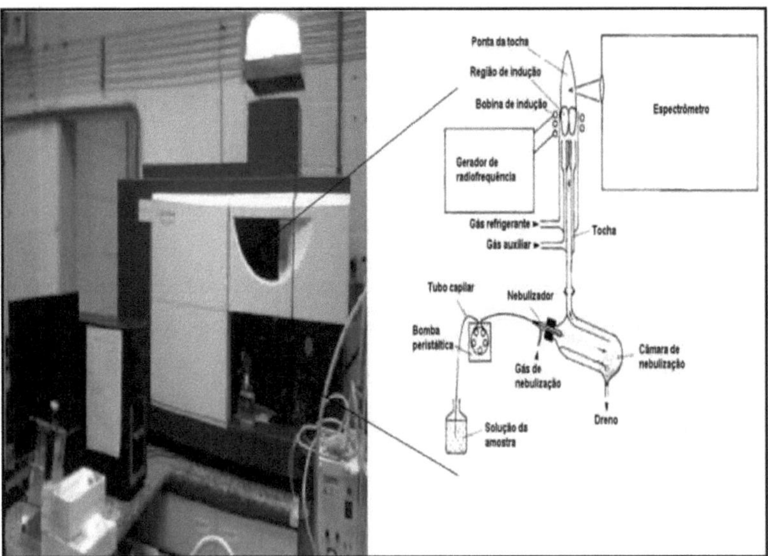

Figure 5 - Agilent 700 Series ICP OES.

To evaluate total metals, 100 mL of the collected water sample was transferred to a beaker and 10 mL of nitric acid (Synth®) was added, then the solution was taken to a hotplate at 100°C and heated until approximately 15 mL of solution (sample + acid) was reached. To prevent the solution from bubbling, glass beads were placed in the beakers.

Before evaluating the dissolved metals, the sample was filtered using a 0.45μm filter (Milipore®). After filtration, 10 mL of nitric acid was added to the 100 mL of filtered sample and it was then taken to a hotplate at 100°C and heated until about 15 mL of solution was reached.

When the solutions reached approximately 15 mL, they were cooled to room temperature and poured into a 25 mL volumetric flask. In this way, the total and dissolved metal samples were preconcentrated four times, and the partial results obtained from the ICP readings were divided by four.

4.3.2 IPMCA calculation

Given the weightings for the variables determined in a water sample, the IPMCA is calculated as follows:

$$IPMCA = PE \times ST$$

(3)

Where:

PE: Value of the highest weighting of the group of essential variables;

ST: Average value of the three highest weights in the group of toxic substances. This value is an integer and the rounding criterion should be as follows: values less than 0.5 will be rounded down and values greater than or equal to 0.5 will be rounded up.

The IPMCA value can vary from 1 to 9 and is subdivided into four quality bands, classifying waters for the protection of aquatic life.

4.3.3 Calculating the EIT

The criterion used to calculate the EIT was in accordance with that proposed for lotic environments by Carlson (1977), modified by Lamparelli (2004) for rivers and streams, represented by the formulas:

$$IET\ (CL) = 10x(6-((-0,7-0,6x(\ln CL))/\ln 2))-20$$

(4)

Where:

CL: chlorophyll-a concentration measured at the water surface, in μg.L-1;

ln: natural logarithm.

IET (Pt) = 10x(6-((0,42-0,36x(ln Pt))/ln 2))-20

(5)

Where:

Pt: total phosphorus concentration measured at the water surface, in µg.L-1 **ln:** natural logarithm.

From the determination of the trophic state indices of these two variables, it is possible to establish an average in relation to the potential and evidence of eutrophication, using the formula:

IET = (IET (CL) + IET (Pt)) / 2

(6)

Where:

IET (CL): determination of the trophic state index for chlorophyll-a;

IET (Pt): determination of the trophic state index for total phosphorus

4.3.4 VAT calculation

VAT should be calculated from IPMCA and IET, according to the expression (CETESB, 2007):

IVA = (IPMCA x 1,2) + IET

(7)

4.3.5 IQA calculation

The IQA is calculated by the weighted product of the water quality corresponding to the nine parameters mentioned above. If the value of any of the nine variables is not available, the calculation of the WQI is unfeasible. Its formula is represented by the following equation (CETESB, 2007):

$$IQA = \prod_{i=1}^{n} q_i^{w_i}$$

(8)

Where:

IQA: Water Quality Index, a number between 0 and 100;

Q_i : quality of the ith parameter, a number between 0 and 100, obtained from the respective "average quality variation curve", as a function of its concentration or measure e,

W_i : weight Corresponding to the i-th parameter, a number between 0 and 1, assigned according to its importance for the overall quality conformation, where:

$$\sum_{i=1}^{n} w_i = 1$$

(9)

Where:

n: number of variables included in the IQA calculation

4.4 Statistical treatment of data

The data was treated according to multivariate statistical analysis, using PCA (principal component analysis) to better evaluate the results of the parameters obtained.

Principal component analysis provides the information needed to understand the role of the original descriptions in forming the principal components. It can also be used to show the relationships between the original descriptions in reduced space. PCA is a way of identifying the relationship between the characteristics of the data extracted by the matrices (LEGENDRE and LEGENDRE, 1998).

A correlation matrix was established and based on this matrix, the PCA was carried out in R mode, using a scalar matrix. The PAST 2.17 software (CHIBA et. al., 2011) was used to prepare the PCA.

5. RESULTS AND DISCUSSION

The results of this research, based on the study of geographic and seasonal points, allowed for a more realistic assessment of the level of quality and information on possible sources of impact. The results will be presented below, with discussions of the parameters analyzed, which were compared with the maximum limits stipulated by CONAMA Resolution 357/05, for class 3 water resources. This class includes the following water uses: supply for human consumption, after conventional or advanced treatment; irrigation of tree, cereal and forage crops; amateur fishing; secondary contact recreation; and animal watering.

5.1 Generation of maps of the study area

The land use (figure 6) and contour lines of the study area (figure 7) are available below.

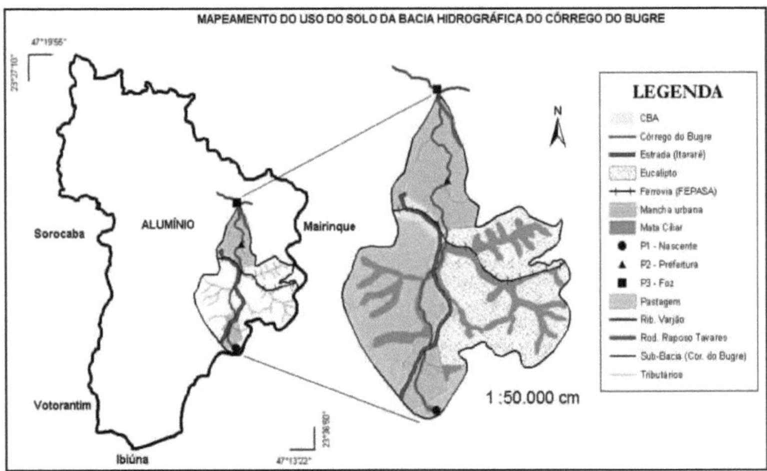

Figure 6 - Land use mapping of the study area.

The Córrego do Bugre has three important springs, all located in the Itararé neighborhood in Aluminio - SP, where the main spring is well preserved. It is approximately 8 km long in a straight line and crosses the entire urban perimeter, in the districts of Itararé, Centro, Vila Paulo Dias and Vila Pedàgio, where it receives untreated domestic effluent and also bauxite and soda residues (diffuse), used in the production of aluminum by CBA (Companhia Brasileira de Aluminio).

The main spring is located at an altitude of 870 m, while the mouth is 740 m, so the Bugre stream has a gradient of 130 meters along its course. Looking at the land use mapping and the contour lines, it is possible to see that basically all of Aluminio's urban densification is concentrated near the headwaters of the stream (Bairro Itararé) and in valley areas, mainly between the 800 and 760 meter contour lines, downstream of the stream. This represents a geographically delicate issue, all the more so given the lack of environmental and urban planning.

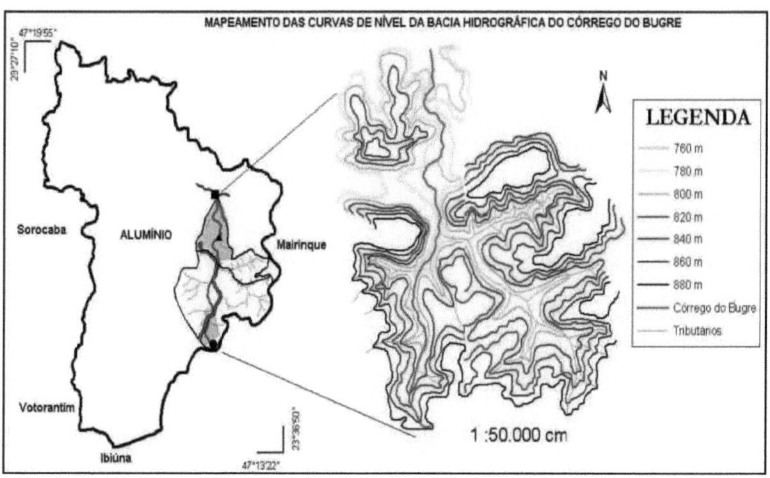

Figure 7 - Mapping of the contour lines in the Bugre stream basin.

This stream began to be polluted in the 1970s, with the increase in industrial activities and the appearance of some sewage discharge points. But it was in the 1980s, with the growth of the urban population and its activities, that there was a large increase in polluting materials in the Bugre stream.

5.2 Rainfall and flows

Below are the results for the average winter and summer flows (table 6). The rainfall for Aluminio in 2013 is shown in figure 8.

Table 6 - Seasonal results of the flow of the Bugre stream

Seasonality	Months	Q (m3/s)
Winter	Jun/13	0,211
	Jul/13	0,194
	Aug/13	0,182
	Average	0,195
	Deviation P.	0,01563
Summer	Dec/13	0,237
	Jan/14	0,279
	Feb/14	0,273
	Average	0,263
	Deviation P.	0,02329

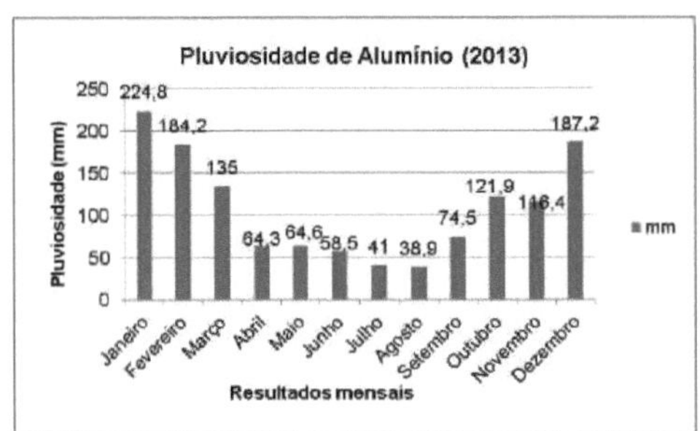

Figure 8 - Average rainfall in Aluminio. Source: CEPAGRI (2013)

Aluminio's annual rainfall is pertinent to the high altitude tropical climate, as the city is situated at an average altitude of over 800m. Rainfall is concentrated in summer, with a maximum annual rainfall of 224.8 mm in January, and drought occurs in winter, with a minimum annual rainfall of 38.9 mm in August (CEPAGRI, 2013).

The flow measurements therefore reflect the annual flood and drought regimes of the Bugre stream (figure 9). When taking into account the summer and winter flow averages, the difference was around Q = 0.068 (m^3 /s), which is equivalent to 68 liters per second. The seasonal average flow for the two seasons was Q = 0.229 (m^3/s). In a possible emergency situation and shortage of water resources in the city of Aluminio, if properly treated, the Bugre stream would contribute 19,785,600 liters of water daily to the public supply of the Sorocaba river basin.

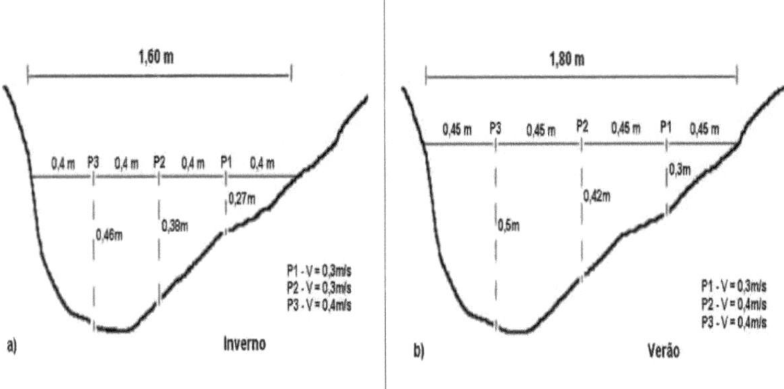

Figure 9 - a) Representation of the measurements determining the flow during the winter period. b) Representation of the average of the measurements determining the flow during the summer period.

According to Rosa, et al. (2012) the processes of a hydrological cycle are influenced by physical (area, topography, vegetation cover, climate, soil) and chemical (industrial and domestic sewage, fertilizers, pesticides, rock alteration) and biological characteristics. The multiple use of water causes changes in its quality, thus influencing the decrease in the availability of water resources, especially in urban regions or regions with intensive industrial and agricultural use. In the world, data on the distribution of water by economic sector shows that agriculture is responsible for 72% of water consumption, followed by industry (22%) and finally domestic use (8%).

5.3 In situ parameters

The analysis of certain parameters *"in situ"* allows the preliminary characterization of the water in a body of water before other characteristics are assessed in the laboratory. The main parameters assessed at the sampling site are hydrogen potential (pH), conductivity, dissolved oxygen and water temperature (ROSA, et. al. 2009). The results for electrical conductivity (figure 10), DO (figure 11), pH (figure 12) and temperature (figure 13) are shown below.

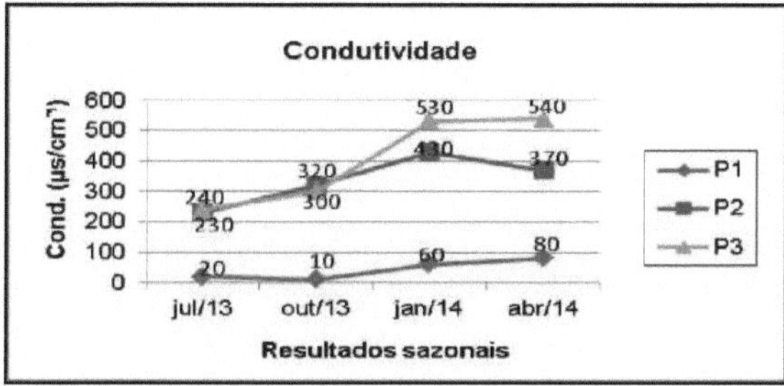

Figure 10 - Seasonal determination of conductivity.

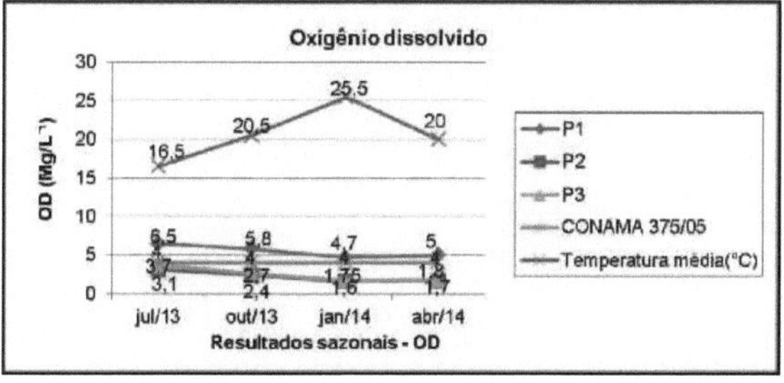

Figure 11 - Seasonal determination of DO.

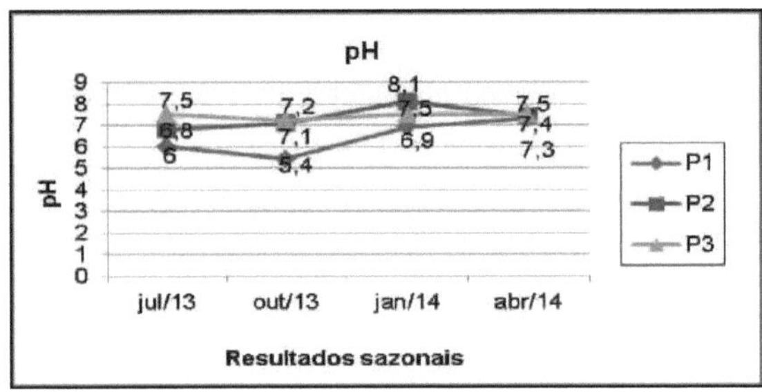

Figure 12 - Seasonal pH determination.

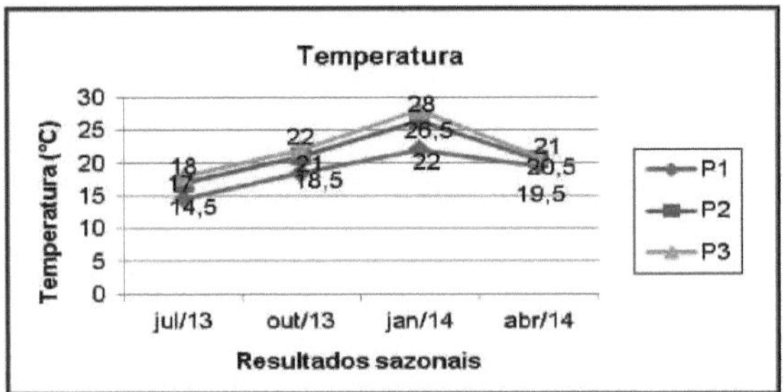

Figure 13 - Seasonal temperature determination.

According to CETESB (2008a), conductivity values above 100 µS cm^{-1} indicate that the body of water is impacted. According to Ribeiro et. al (2004), electrical conductivity is the variable most commonly used to assess the level of salinity, or the concentration of soluble salts in irrigation water and soil. This measure increases proportionally as the concentration of salts increases. Points 2 and 3 showed very high conductivity values, especially in summer, due to the increase in effluent discharge in urban areas. The spring (P1) showed a large increase in conductivity levels, however it was the only point to have values below 100µs⁄cm .$^{-1}$

During the summer period, the DO was significantly lower than in winter. This is because the increase in temperature contributes to the proliferation of microorganisms such as fecal coliforms. Many of these micro-organisms (aerobes) consume the oxygen dissolved in the water during the respiration process, which increases the $BOD5_{,20}$ especially in warmer periods. In relation to CONANA Resolution 357/05, points 2 and 3 showed critical levels, even for class 3 watercourses. During the periods from January to April, the levels of DO were below 4 mg/L ″l, the established minimum level.

As for pH, only the October 2013 sample had a value below 6 (5.4), showing moderate acidity. In the other samples, the pH remained in the range of 6 to 9 in all readings and for all points, even at points 2 and 3 where the Bugre stream is intensely impacted. Biological systems are very sensitive to pH values and federal legislation sets pH values between 6.0 and 9.0 as a criterion for protecting aquatic life (CONAMA, 2005). According to Calmano et al. (1993), redox potential and pH affect the mobility of metals from the sediment to the water column, where pH is the main mobilization factor.

The temperatures showed the expected oscillations according to seasonality, but P3 showed an amplitude of 10°C between the Jul/13 and Apr/14 analyses. Bello and Guandique (2011) obtained similar levels of amplitude between the months of July and March, where P2 showed 10.9°C and P3 10.1°C respectively for the Ipanema River. Most biological processes occur more intensely with this increase in temperature, causing an acceleration in metabolism and an increase in the need for oxygen, which leads to increased competition and serious problems due to the low affinity of hemoglobin with heated oxygen (DAVIS & CORNWELL, 1998; PEREIRA, 2004).

When studying the waters of the Sorocamirim River, one of the main sources of the Sorocaba River, Oliveira (2012) obtained more satisfactory levels of some of the parameters discussed above, such as conductivity - maximum: 90 µs⁄cm^{-1} and minimum: 66 µs⁄cm^{-1} and DO - maximum: 6.8 mg/L "' and minimum: 3.7 mg/L "I,

5.4 Metals.

The values of the correlation coefficient R^2 (table 7), the limits of detection and quantification of dissolved metals (table 8) and total metals (table 9) are shown below.

Table 7 - seasonal values of R^2 for the metals obtained

Seasonal results of R^2 for metals (%)

	Jul/13	Oct/13	Jan/14	Apr/14
Cr	99,95	99,95	99,98	99,98
Cu	99,97	99,97	99,99	99,96
Ni	99,94	99,98	99,97	99,99
Pb	99,93	99,97	99,98	99,98
Zn	99,96	99,99	99,97	99,99

Table 8 - Seasonal results of the limits of detection (LOD) and quantification (LQ) for dissolved metals

	Jul/13	Oct/13	Jan/14	Apr/14	
LD	0,0036	0,0046	0,0027	0,0052	Cr
LQ	0,0099	0,0108	0,0087	0,0127	
LD	0,0017	0,0012	0,0008	0,0032	Cu
LQ	0,0056	0,0031	0,0025	0,0078	
LD	0,0009	0,0041	0,0025	0,0017	Ni
LQ	0,0023	0,0102	0,0058	0,0043	
LD	0,0060	0,0065	0,0015	0,0030	Pb
LQ	0,0142	0,0147	0,0048	0,0084	

LD	0,0925	0,0195	0,0037	0,0098	Zn
LQ	0,2206	0,0431	0,0105	0,0274	

Table 9 - Seasonal results of the limits of detection (LOD) and quantification (LQ) for total metals

	Jul/13	Oct/13	Jan/14	Apr/14	
LD	0,0136	0,0057	0,0048	0,0089	Cr
LQ	0,0320	0,0216	0,0170	0,0334	
LD	0,0053	0,0078	0,0026	0,0037	Cu
LQ	0,0142	0,0194	0,0074	0,0106	
LD	0,0059	0,0053	0,0019	0,0044	Ni
LQ	0,0155	0,0147	0,0054	0,0116	
LD	0,0122	0,0124	0,0105	0,0127	Pb
LQ	0,0228	0,0244	0,0373	0,0395	
LD	0,2401	0,0464	0,0287	0,1955	Zn
LQ	0,4870	0,0954	0,0712	0,6816	

Determining the method's limit of detection (LOD) represents the lowest concentration of the analyte under examination in the active ingredient: matrix ratio that can be detected with a certain reliability using a given experimental procedure. Determining the limit of quantification (LQ) represents the lowest concentration that can be identified and quantified in a given matrix with a certain limit of reliability, usually between 95% and 99% (SCHARTZ and KRULL, 1998).

Seasonality was very important in determining the limits of detection and quantification of the metals analyzed in the study area, since in dry seasons the limits were higher and in rainy seasons they were lower.

Graphs showing the results of total and dissolved metals are available below. The discussion follows the presentation of the results.

a. Cadmium

No cadmium was found in any of the samples taken

b. Lead

The values for total lead (figure 14) and dissolved lead (figure 15) are shown below.

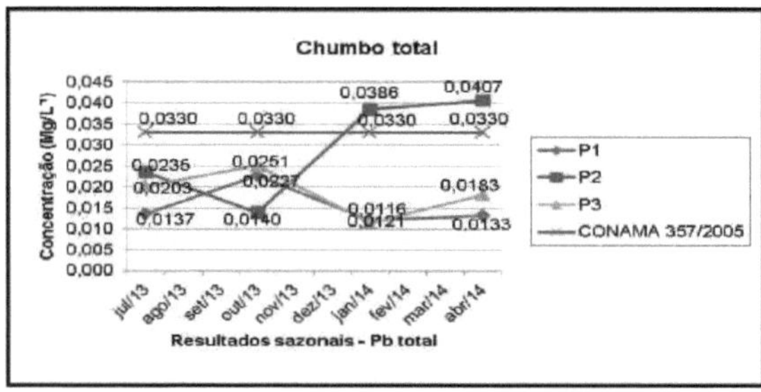

Figure 14 - Seasonal results of total Pb.

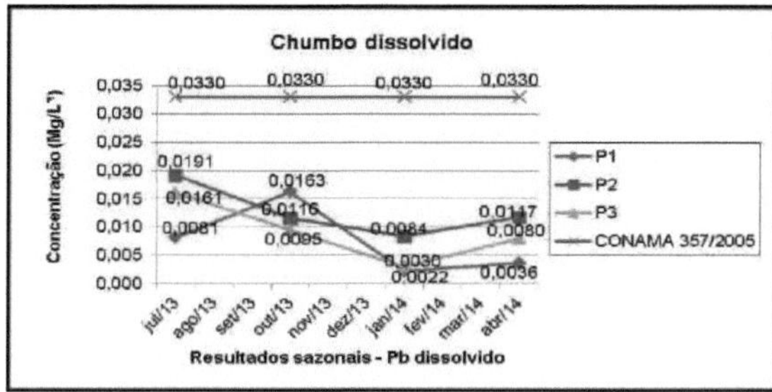

Figure 15 - Seasonal results of dissolved Pb.

c. Copper

The total (figure 16) and dissolved (figure 17) copper values are shown below.

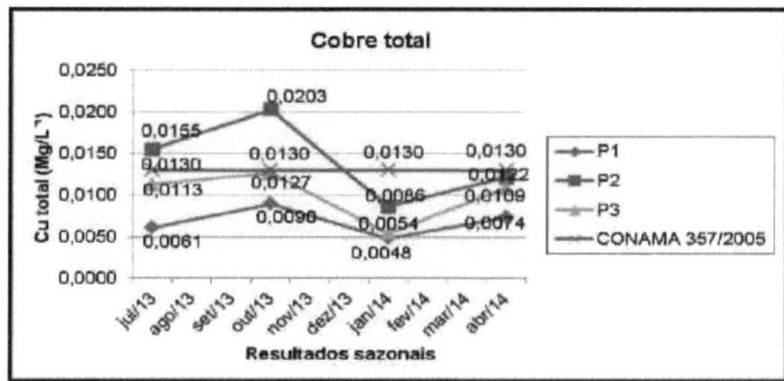

Figure 16 - Seasonal results for total Cu.

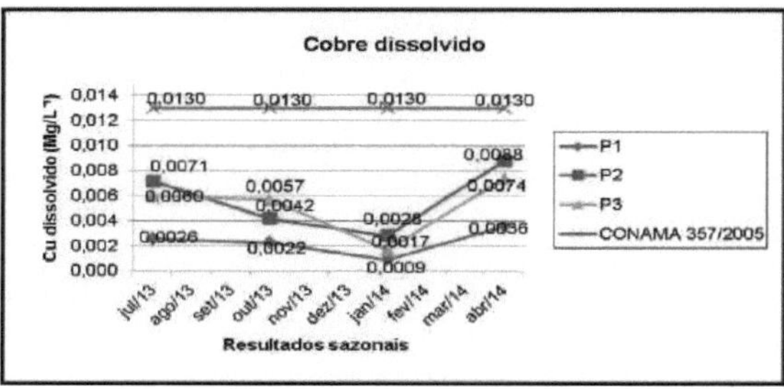

Figure 17 - Seasonal results of dissolved Cu.

d. Chrome

The total (figure 18) and dissolved (figure 19) chromium values are shown below.

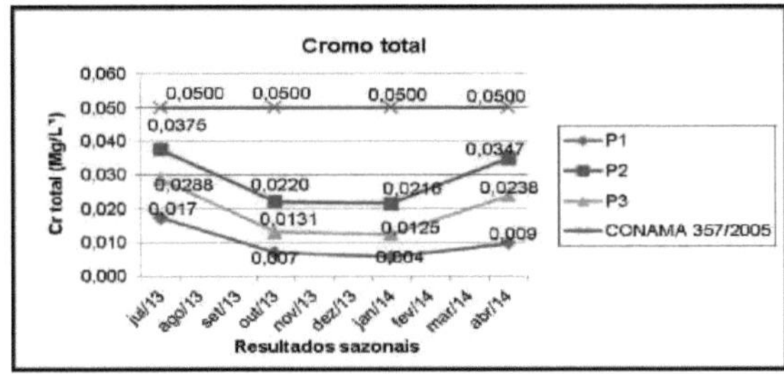

Figure 18 - Seasonal results for total Cr.

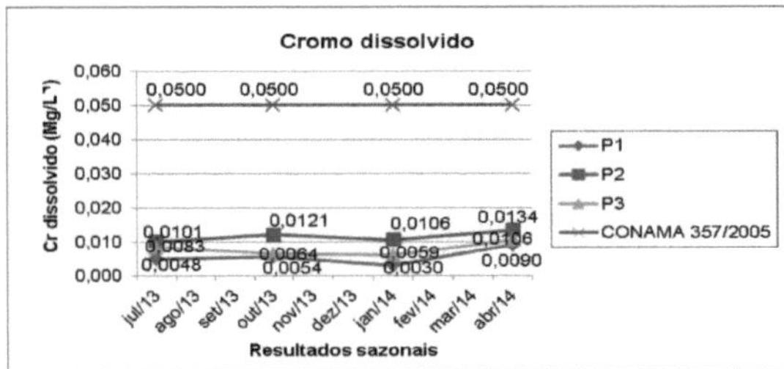

Figure 19 - Seasonal results of dissolved Cr.

e. Mercury

No mercury was found in any of the samples taken.

f. Nickel

The total (figure 20) and dissolved (figure 21) nickel values are shown below.

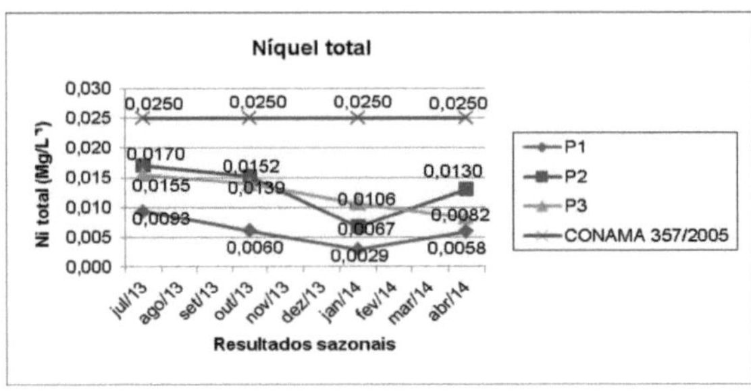

Figure 20 - Seasonal results for total Ni.

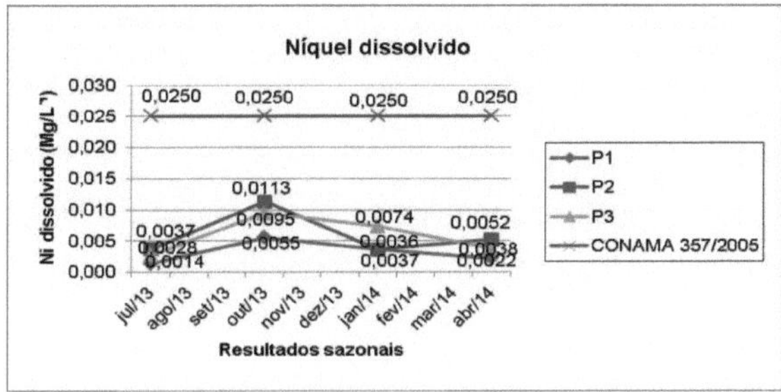

Figure 21 - Seasonal results of dissolved Ni.

g. Zinc

The total (figure 22) and dissolved (figure 23) zinc values are shown below.

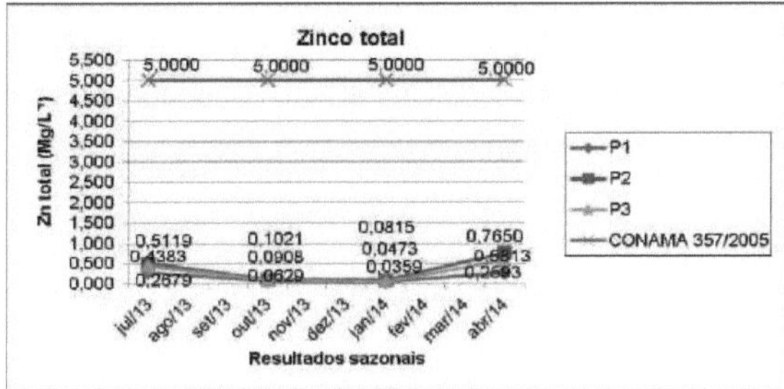

Figure 22 - Seasonal results of total Zn.

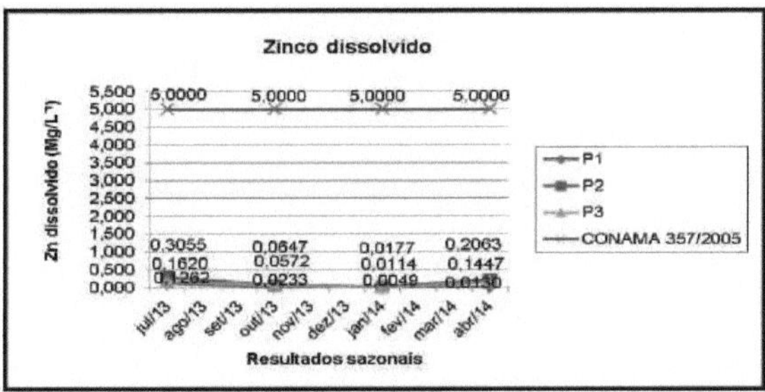

Figure 23 - Seasonal results of dissolved Zn.

P2 stood out for its concentration of total lead in the January and April 2014 samples and total copper in the July and October 2013 samples, with values above those established by CONAMA Resolution 357/2005. All the points in all the collections showed results for chromium, nickel and zinc, as well as all the dissolved metals analyzed below the limits established by CONAMA Resolution 357/2005.

The analysis of metals in the Monjolino and Feijão rivers in São Carlos, SP, showed high levels of Cr, Ni, Pb and Cd in the rainy season, while the dry season showed high levels of Zn and Ni (CHIBA et. al, 2011).

In the aquatic environment, heavy metals can occur in various forms: in solution in ionic form or in the form of soluble organic or inorganic complexes; forming or being retained in mineral or organic colloidal particles; being retained in the sediment; or incorporated into the biota (FEEMA, 1992).

According to Baird (2002), the toxicity of a metal in water varies according to pH and the levels of dissolved and suspended carbon, since metals interact with carbon and its compounds, forming complexes or being adsorbed. The most toxic form of a metal is not free, but when it is found as a cation or bound to carbon chains. In organisms, the main mechanism of toxic action of metals stems from their affinity for sulphur. Thus, when present in their cationic forms, metals react with the sulfhydryl radical (-SH) present in the protein structure of enzymes, altering their properties, which can result in harmful consequences for the metabolism of living beings, such as mercury and lead.

5.5 Phenols and surfactants

The results for total phenols and surfactants are shown in Figures 24 and 25 respectively.

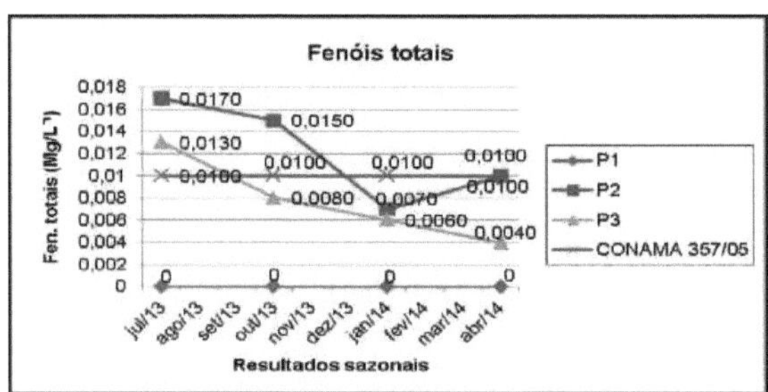

Figure 24 - Seasonal determination of phenols.

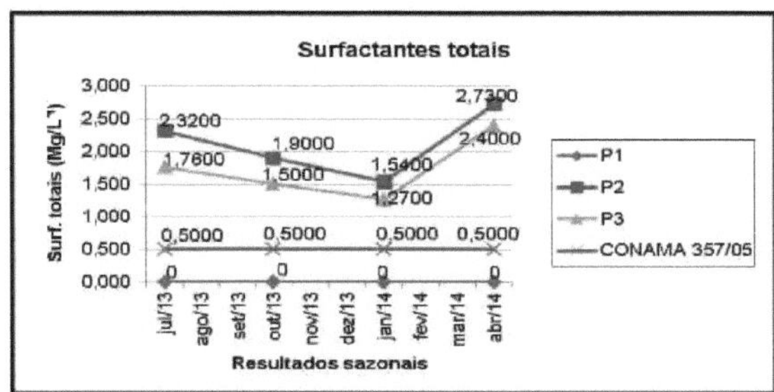

Figure 25 - Seasonal determination of surfactants.

P2 and P3 showed levels above those established by CONAMA Resolution 357/2005, which stipulates a limit of 0.01 mg/L for phenols and 0.5 mg/L for surfactants, respectively. P2 and P3, due to the dumping of untreated domestic sewage, showed very high levels of surfactants in all the periods analyzed. With regard to phenols, it is most likely that their presence in the Bugre stream was due to diffuse pollution. Thus, there was a significant influence of land use and seasonality on the concentration of these substances, with greater availability in the winter period and falls in their concentrations in the summer, the season in which the rainy season occurs in Aluminio.

The current disposal of detergents in sewage causes surfactant levels to vary from 1 to 20 mg/L. The excessive presence of surfactants in the aqueous environment can cause serious damage to aquatic life and water quality, as well as complications in treatment plants. Phenols, when present in treatment plants (water or sewage) react with chlorine to form the compound chlorophenol, which has an unpleasant odor (GARCEZ, 2004), and can also form potentially carcinogenic compounds such as pentachlorophenol (ROSA, et. al 2012).

5.6 Physico-chemical and biological parameters

The seasonal results for chlorophyll-a (figure 26), fecal coliforms (table 10), BOD (figure 27), total phosphorus (figure 28), total nitrogen (figure 29), total solids (figure 30) and turbidity (figure 31) are shown below.

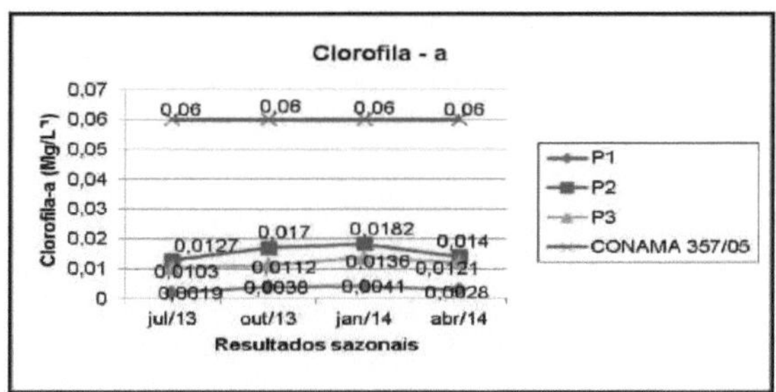

Figure 26 - Seasonal determination of chlorophyll-a.

Table 10 - Seasonal fecal coliform results

Fecal coliforms (MPN/100 ml)				
Points	Jul/13	Oct/13	Jan/14	Apr/14
P1	7	21	79	80
P2	>160000	>160000	>160000	>160000
P3	50.000	54.000	>160000	>160000

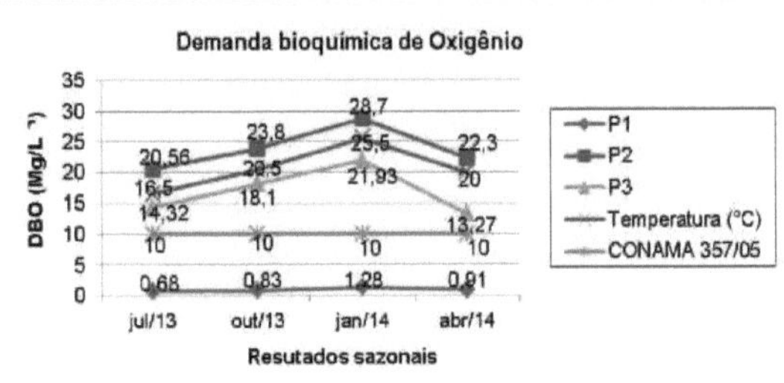

Figure 27 - Seasonal BOD determination.

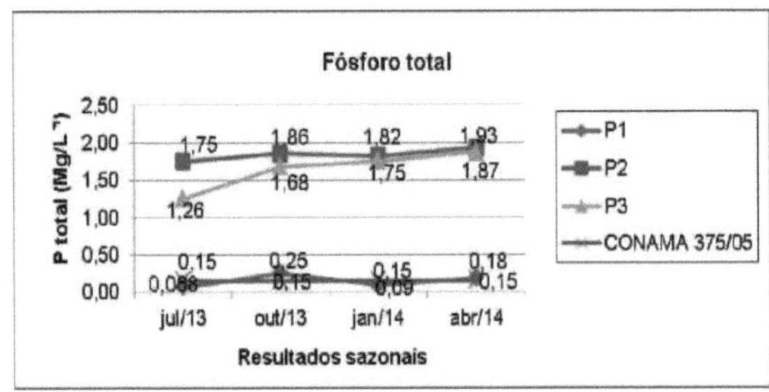

Figure 28 - Seasonal determination of total phosphorus by sampling points.

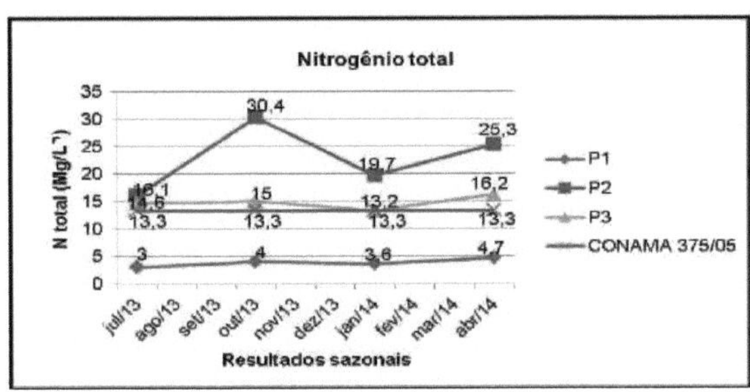

Figura 29 - Seasonal determination of total phosphorus by sampling points.

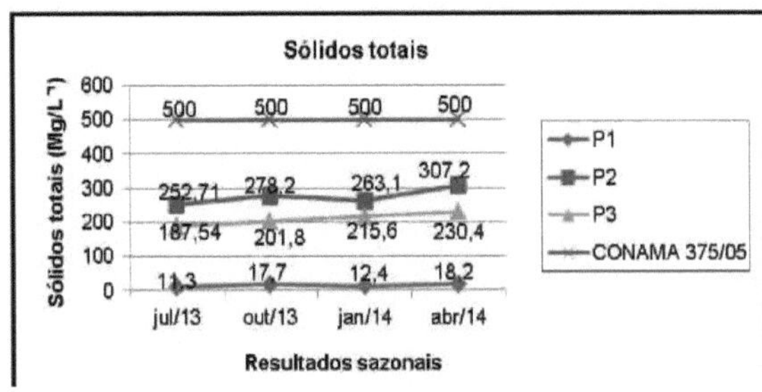

Figura 30 - Seasonal determination of total solids by sampling points.

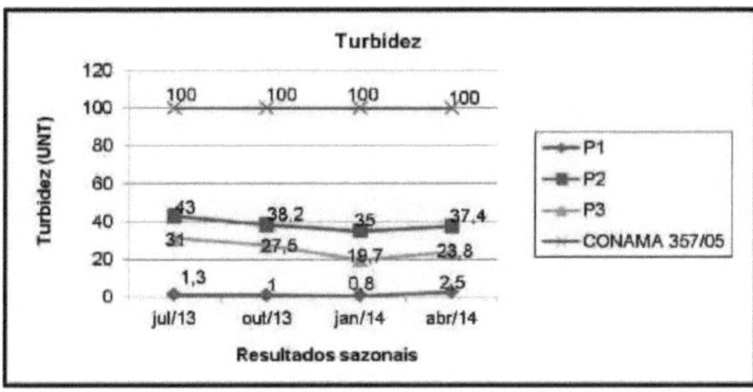

Figure 31 - Seasonal determination of turbidity by sampling points.

The chlorophyll-a results were below the limit stipulated by CONAMA Resolution 357/2007 of 60 µg/L for class 3 watercourses. With regard to fecal coliforms, according to CONAMA Resolution 357/2005, the maximum number for class 3 watercourses is 4000 thermotolerant coliforms per 100 mL. Therefore, the samples from P2 and P3 showed levels of fecal coliforms well above those permitted for class 3 watercourses. In all the samples from points 2 and 3, $BOD5_{.20}$ was above the limits established by CONAMA Resolution 357/2005, which stipulates 10 mg/L for class 3 watercourses. In general, the differences in temperature between summer and winter had a significant impact on seasonal variations in fecal coliforms, BOD and chlorophyll-a.

At points 2 and 3, the phosphorus and nitrogen variables showed high concentrations due to the dumping of untreated domestic effluent in the urban perimeter, with levels much higher than those established by CONAMA Resolution 357/05 for class 3 water resources. The October/2013 and April/2014 nitrogen samples from P2 needed to be diluted by 50%, so both exceeded the detection limits of the methodology used.

All the sampling points showed total solids and turbidity values lower than those established by CONAMA Resolution 357/2005, which stipulates 500 mg/L of total solids and 100 UNT of turbidity respectively as maximum limits for class 3 watercourses. Sampaio et. al (2007) obtained a linear relationship for conductivity as a function of total solids in pig and dairy wastewater, less so for domestic wastewater.

5.7 Determination of EIT, IPMCA, IQA and VAT

The results for the EIT (figure 32), IPMCA (figure 33), IQA (figure 34) and IVA (figure 35) are available below.

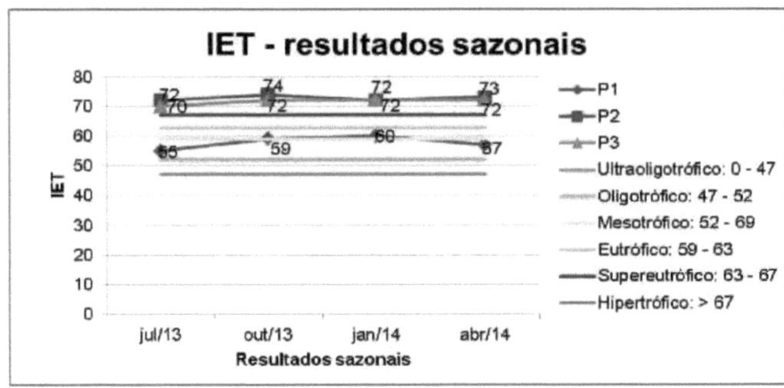

Figure 32 - EIT seasonal averages.

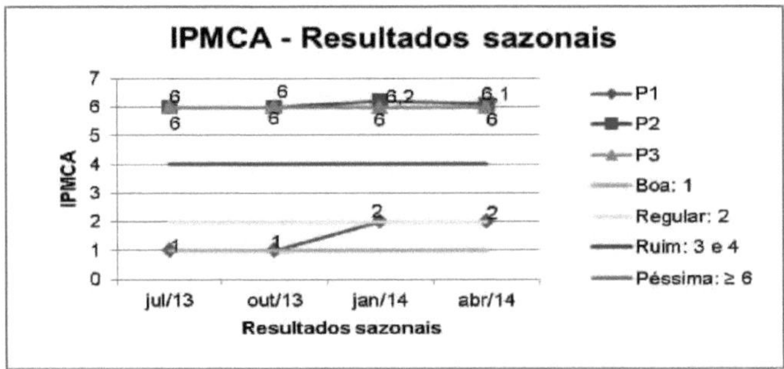

Figure 33 - IPMCA seasonal averages.

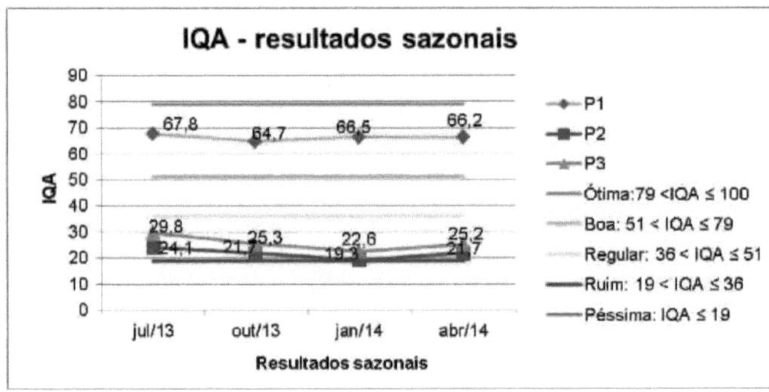

Figure 34 - WQI seasonal averages.

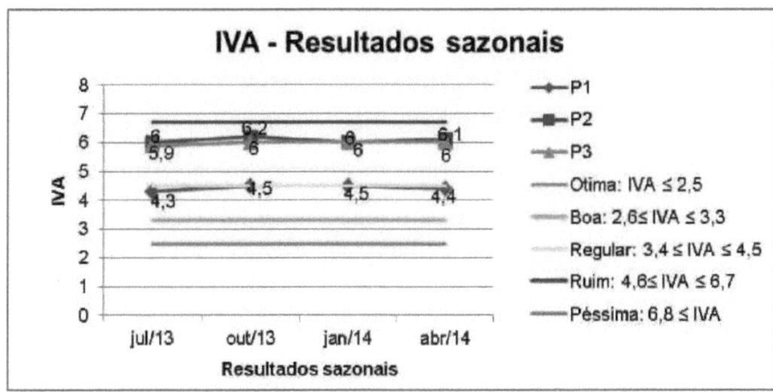

Figure 35 - VAT seasonal averages.

The seasonal variation in the EIT was not intense between points 2 and 3, which are hypereutrophic. However, at P1, seasonality was responsible for the change in category from mesotrophic (July and October 13) to eutrophic (January 14). Bello and Guandique (2011) obtained similar EIT levels between 45.9 (April) and 72.1 (December) in 2009 in the Ipanema River, an important tributary of the Sorocaba River. In a lake located in Carlos Alberto de Souza Park in the Campolim district of Sorocaba, Gouveia et. al (2014) showed that its water is hypereutrophic, with an EIT of 72.98.

At points 2 and 3, after April 2014, nitrogen was considered the limiting agent for algae growth after the averages of the correlation analyses with phosphorus (Table 11), as the nitrogen/phosphorus ratio (N:P) was less than 12. Phosphorus was considered the limiting agent in the correlation with nitrogen in all the periods analyzed in P1, where the N:P ratio was greater than 12. When the N:P ratio is greater than 12, the system indicates an excess of nitrogen and a deficiency of phosphorus (NASELLI-FLORES, 1999).

Table 11 - Seasonal values of the N:P ratio

Samples	Jul/13	Oct/13	Jan/14	Apr/14
P1	138,3	32	16	40
P2	15,24	9,48	10,5	10,82
P3	17,7	11,42	8,98	7,54

Points 2 and 3 showed very poor water quality in terms of IPMCA in all the samples. The seasonal variations in the IPMCA at P1 stand out. According to the metals results, there were high concentrations of lead in the samples collected in January and April 2014.

P1 had a good WQI in all the analyses carried out, with a minimum value of 64.7 in Oct/13 and a maximum of 67.8 in Jul/13. Points 2 and 3 had poor WQIs in all the analyses. Bello and Guandique (2011) obtained WQI levels between 40 and 80, where the highest levels were in the months of low rainfall (July 78 and August 79, respectively). Despite being impacted by high eutrophication, Gouveia et. al (2014) obtained a good WQI (65.11) in the lake analyzed in Sorocaba.

The VAT, an index made up of the EIT and IPMCA, remained regular in all the analyses carried out in P1, remaining at the maximum limit for the bad category in the months of October (2013) and January (2014). P2 and P3 presented this bad index. Seasonality was not very decisive in the evaluation of this index, where the main factor influencing changes in categories between the points was land use.

5.8 Statistical analysis of the results.

The graphs below show the results of the principal component analysis for the indices and parameters evaluated. The results of the PCA for the evaluation of the EIT (figure 36), IQA (figure 37), IPMCA (figure 38) and physico-chemical parameters (figure 39) are shown below.

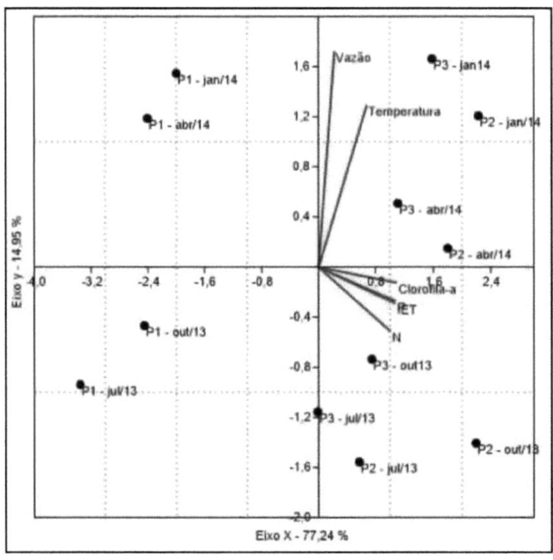

Figure 36 - Seasonal PCA results for the EIT components.

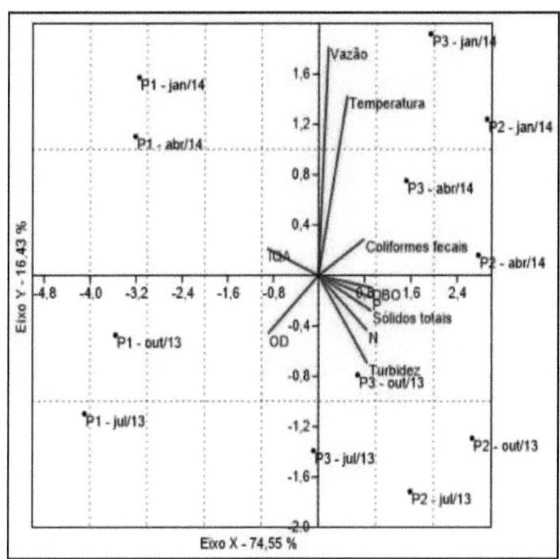

Figure 37 - Seasonal results of the PCA for the WQI components.

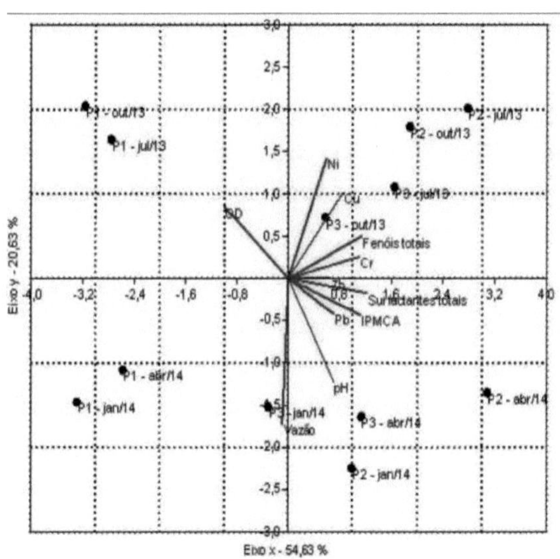

Figure 38 - Seasonal PCA results for the IPMCA components.

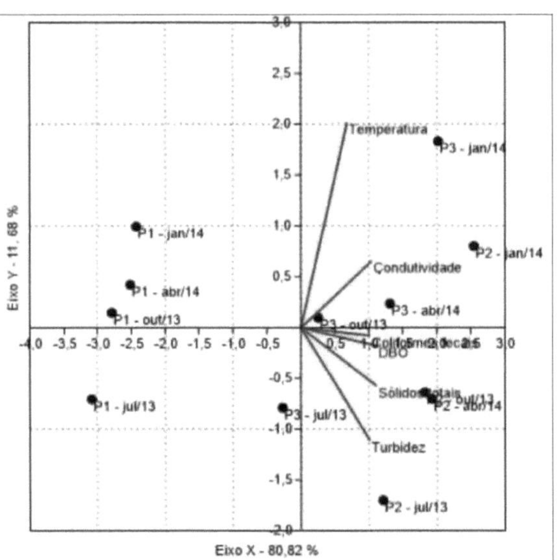

Figure 39 - Seasonal PCA results for physico-chemical parameters.

In the PCA graphs of the components of the EIT, IQA, IPMCA and physico-chemical parameters, axis 1, which is mainly made up of the relationship between the parameters and indices and the different sampling points, explained more than 50% of the analyses carried out in all cases. The percentage was over 70% in the case of the EIT and IQA, and over 80% in the case of the physico-chemical parameters, which showed the great influence of land use on the samples taken.

The high levels of WQI were more related to the amount of DO present in the stream, justifying P1 as the best point for the Water Quality Index in all the samples. P2 and P3 had poor WQIs in all evaluations, mainly due to the large presence of fecal coliforms, high BOD (above 10 mg/L), DO below 4 mg/L, as well as total phosphorus and nitrogen concentrations well above those permitted for class 3 watercourses. Fluctuations in the EIT were closely related to fluctuations in nitrogen and chlorophyll-a, but especially phosphorus. Seasonality was a determining factor only for the variation in EIT at P1, a point that is little impacted.

The IPMCA, according to the principal component analysis, showed a very close correlation with surfactant levels in the Bugre stream, which were greatly influenced by the seasonality of their concentrations due to the increased volume of water in warmer seasons. Regarding the relationship between metals and flow, according to Chiba et. al (2011) point metal contamination (industrial waste, for example) is more evident during the dry season. This type of contamination provides a constant supply of metals in the water. During rainy periods, the levels of metals in water bodies should be lower than the levels of metals during dry periods, due to the decrease in the concentration of metals due to the increase in the flow of rivers and streams.

P2 had the worst water quality indices because it was closest to the effluent discharge points in the urban perimeter.

The results showed that seasonality and land use were the two determining factors for the variations in the indices evaluated. However, land use was decisive for water quality at all points (figure 39) and at all times of the year, demonstrating that continuous characteristics (such as land use) were more important than temporary characteristics (seasonality) in determining water quality.

Figure 40 - Mapping the water quality of the Bugre stream

The development of environmental zoning and management allowed for the development of a more complex environmental analysis of the Bugre stream basin. The figure shows that the water quality was generally satisfactory from its source (upstream) through the Itararé district to the main urban concentration in Aluminio, where untreated domestic effluent is discharged downstream. According to Prado et al. (2003), population growth is one of the main causes of water quality degradation *in* rivers and reservoirs, since there is a proportional increase in the production of liquid waste, most of which is discharged into water bodies.

The existence of complex aquatic life (such as fish) was not verified in the Bugre stream in all the samples taken at points 2 and 3, a fact proven by the results of the indices presented. In this context, no toxicity tests were carried out. According to the IPT (2006), the main problems faced by UGRHI 10 are, among others, the lack of data and studies on the resources in this region, lack of adequate sewage treatment, losses in supply systems, lack of conservation measures and protection of water sources, impairment of water bodies, eutrophication of water resources and fish mortality.

In view of the above, even if Aluminio does not have a Municipal Master Plan because it does not yet have 20,000 inhabitants, it is pertinent (even as a strategic issue for the future of the municipality) that the public authorities of this municipality consider the data from this work regarding the environmental diagnosis of the Bugre stream basin.

For Zacharias (2006), one of the hypotheses of his research into the territorial zoning of Ourinhos suggests that environmental zoning work will only be truly effective if it goes beyond and aligns with the legislation and guidelines of the Municipal Master Plan, as it represents the major Organic Law, which governs the City Statute in favor of sustainable development, with environmental policies and management.

It is important to emphasize that in this case the data from this research will only have any public and social relevance for the municipality of Aluminio if there is effective participation by the public authorities and society in making decisions that touch on the issues raised here.

6. CONCLUSIONS

The tools and methods used were relevant to a complex assessment of the study area. In view of the evaluation of the indices and parameters analyzed, it is possible to infer that the water of the Córrego do Bugre has been showing unsatisfactory quality indices for class 3 watercourses downstream (where the greatest urban concentration occurs). According to the methodologies applied, land use was more decisive than seasonality, since land use represented continuous characteristics and seasonality, temporary characteristics of the water parameters. The dumping of domestic and industrial effluents in urban areas has caused the water quality of the Bugre stream to decline sharply in this stretch.

As for seasonality, determining the flow rate was very important for assessing the volume of water in the Bugre stream during the dry season (winter) and the flood season (summer), as there were variations in the concentration of pollutants at different times of the year. The spring, being relatively well preserved, showed significantly higher water quality in terms of the indices and parameters evaluated. As this is an important source of water in the Sorocaba river basin, and a strategic area for collecting water for public supply, it is necessary to implement measures to reduce the anthropogenic impact caused mainly in the urban area, such as the treatment of domestic effluents mainly downstream of the Bugre stream.

Among other measures, the population living near the stream should also be involved to help them understand and identify any points that need to be analyzed. This will increase the accuracy of the diagnosis and, consequently, enable more appropriate actions to be taken in response to the problems identified.

7. BIBLIOGRAPHICAL REFERENCES

AISSE, M. M. **Sistemas Econômicos de Tratamento de Esgotos Sanitários.** Rio de Janeiro: ABES, p. 192, 2000.

NATIONAL WATER AGENCY - ANA. **National Guide to** the **Collection and Preservation of Samples**: Water, sediments, aquatic communities and liquid effluents. Ed. Athalaya. 2011.

ALLOWAY, B.J. & AYRES, D.C. **Chemical Principles of Environmental Pollution**, 2 ed., Chapman & Hall; New York; 1997.

AMERICAN PUBLIC HEALTH ASSOCIATION - APHA. **Standard methods for the examination of water and watwater, 21sted**. Washington. 2005.

ANDREWS, J. E.; BRIMBLECOMBE, P.; JICKELLS, T. D.; LISS, P. S.; REID, B. J. **An introduction to environmental chemistry**. 2th ed. Blackwell Publishing. p. 296. 2004.

ARAÙJO, R.; GOEDERT, W. J.; LACERDA, M. P. C. **Quality of a soil under different uses and under native savannah.** Revista Brasileira de Ciência do Solo. p. 31:1099 - 1108. 2007.

BAIRD, C. **Environmental chemistry**. Trad. RECIO, M. A. L.; CARRERA, L. C. M. 2. ed. Porto Alegre: Bookman, 2002.

BALLS, P. W.; BROCKIE, N.; DOBSON, J.; JONHSTON, W. Dissolved oxygen and nitrification in the upper forth estuary during summer (1982-92): *patterns and trends*. **Estuarine, coastal and shelf science**. v.42, p. 117- 134. 1996.

BARRA, C. M. et al. **Arsenic specification - a review.** Quimica Nova, v. 23, n. 1, p. 58-70, 2000.

BELLO, F. H.; GUANDIQUE, M. E. G. **Environmental diagnosis of the aquatic environment of the Ipanema River, Sorocaba/SP.** Holos Environment, v. 11, n. 2, p. 94. 2011.

BIGARDI. **Fate of anionic surfactants in an aerated lagoon sedimentation pond WWTP**. Revista Engenharia Sanitària e Ambiental. v.8, p. 45-48. 2003.

BRAGA, B.; HESPANHOL, I.; CONEJO, J. G. L.; MIERZWA, J. C.; BARROS, M. T. L. de; SPENCER, M.; PORTO, M.; NUCCI, N.; JULIANO, N.; EIGER, S. **Introdução à engenharia ambiental.** São Paulo: Pearson Prentice Hall. p. 306. 2005.

CACCIA, V.G., MILLERO, F.J., PALANQUES, A. The distribution of trace metals in Florida

Bay sediments. **Marine Pollution Bulletin** v. 4611, p. 1420-1433. 2003

CALMANO, W.; HONG, J.; FORSTNER, U. **Binding and mobilization of heavy metals in contaminated sediments affected by pH and redox potential**. v. 28 (8-9), p. 53-58. 1993.

CÂMARA, G.; MEDEIROS, J., S. **Geoprocessing for environmental projects.** 2.ed. Sâo José dos Campos, 2001. Available at:<http://www.dpi.inpe.br/gilberto/livro/introd/cap10aplicacoesambientais.pdf> Accessed on: May 31, 2013.

CARLSON, R. E. A trophic state index for lakes. **Limnol. and Oceanogr.,** v.2, n.2, p. 361-369. 1977.

CARVALHO, T. M. **Conventional and unconventional flow measurement techniques.** RBGF - Brazilian Journal of Physical Geography. Recife, PE. Vol. 01 n.01 May/Aug, p. 73-85, 2008.

CEPAGRI - **Climate of São Paulo municipalities**. 2013. Available at: <http://www.cpa.unicamp.br/outras-informacoes/clima-dos municipiospaulistas.html> Accessed on: April 10, 2014.

CHIBA, W. A. C.; PASSERINI, M. D.; BAIO, J. A. F.; TORRES, J. C.; TUNDISI, J. G. **Seasonal study of contamination by metal in water and sediment in a sub-basin in the southeast of Brazil**. Braz. J. Biol. v.71, n. 4. Sao Carlos, SP. 2011.

CODE PERMANENT: ENVIRONNEMENT ET NUISANCES. **Legislative and administrative editions**. Paris, France. Vol.1, 2. ed, p. 1784. 1986.

COMPANHIA DE TECNOLOGIA DE SANEAMENTO AMBIENTAL - CETESB. (2008a). **Water quality variables**. Available at: <http://www.cetesb.sp.gov.br/Agua/rios/variaveis.asp>, Accessed on: April 16, 2014.

CoMPANHIA DE TECNoLoGIA DE SANEAMENTo AMBIENTAL - CETESB. **indice de Qualidade das Aguas Interiores Do estado de Sao Paulo - Annex III: indice de Qualidade das Aguas**. Sao Paulo, SP. 2007. Available at: <http://pt.scribd.com/doc/32168020/Anexo-III#download> Accessed on: 05 Apr. 2013.

ENVIRONMENTAL SANITATION TECHNOLOGY COMPANY - CETESB. **Water quality variables**. 2008. Available at:

<http://www.cetesb.sp.gov.br/Agua/rios/variaveis.asp>, Accessed on: April 20, 2013.

NATIONAL ENVIRONMENTAL COUNCIL - CONAMA. Provides for the classification of bodies of water and environmental guidelines for their classification, as well as establishing the conditions and standards for effluent discharges, and other measures. Resolution n. 357 of March 17, 2005. **CONAMA Resolutions**: resolutions in force published between July 1984 and November 2008. Brasilia, 2. ed., p. 928. 2008.

CRUZEIRO DO SUL. **Sabesp plans to have treated sewage by May 2015.** Available at: <http://www.cruzeirodosul.inf.br/materia/534278/sabesp-planeja-ter- esgoto-tratado-ate-maio-de-2015> Accessed on: Nov. 10, 2014.

CULLEN, J. J. The deep Chlorophyll maximum: Comparing vertical profiles of chlorophyll a. **Aquatic Science** v. 39, p.791-803. 1982.

DAVIS, M. L; CORNWELL, D. A. **Introduction to environmental engineering**. 3rd ed., Singapore: McGraw-Hill, p. 917. 1998

EGWARI, L.; ABOABA, O. O. Environmental impact on the bacteriological quality of domestic water supplies in Lagos, Nigeria. **Revista Saùde Pùblica**, v.36, p. 513-520. 2002.

FEEMA; **Water pollution of Guanabara Bay by heavy metals chromium and zinc**. Rio de Janeiro, Brazil, 1992.

FORESTRY FOUNDATION. **APA** Itupararanga- **Management Plan Available** at:

<http://www.fflorestal.sp.gov.br/media/uploads/planosmanejo/APAitupararanga/PM_%20APA_Itup_final.pdf> Accessed on May 19, 2012.

BRAZILIAN INSTITUTE OF GEOGRAPHY AND STATISTICS - IBGE. **Statistical Municipal Map of Aluminio - 2010 Census.** Available at: <ftp://geoftp.ibge.gov.br/mapas_estatisticos/censo_2010/mapa_municipal_estatistico/sp/aluminio_v2.pdf>

BRAZILIAN INSTITUTE OF GEOGRAPHY AND STATISTICS - IBGE. **Population of Aluminio, SP - 2010 Census.** Available at: <http://www.ibge.gov.br/cidadesat/topwindow.htm?1> Accessed on: 06 Apr. 2012.

GARCEZ, L. N. **Manual of laboratory procedures and techniques for the analysis of sanitary and industrial water and sewage**. USP (Department of Hydraulic and Sanitary Engineering - Sanitation Laboratory), p. 105, 2004.

GOMES, J. & SANTOS, I. Comparative analysis of liquid discharge measurement campaigns using conventional and acoustic methods. In: **XIII Simpòsio Brasileiro de Recursos Hidricos**, p.32-36. 2003.

GOVEIA, D.; REBELO, A.; LORO, A. P.; SASSO, G. D.; DA ROCHA T. N. F.; DOMPIERI, T. P.; CARLOS, V. M. **Use of quality indices to evaluate water in a lentic environment.**

Brazilian Journal of Biosystems Engineering v. 8(2): 104-111. 2014.

INSTITUTE OF TECHNOLOGICAL RESEARCH - IPT. **Technical Report No. 91 265-205**. USP. Sao Paulo, SP. 2006.

INSTITUTE OF TECHNOLOGICAL RESEARCH - IPT. **Geomorphological Map of the State of Sao Paulo**. V.1. Publication 1183. 1981.

JONNALAGADDA, S.B.; MHERE, G. Water Quality of the Odzi river in the eastern highlands of Zimbabwe. **Water Research**, v. 35, p. 2371-2376. 2001.

JULIO, M.; FILHO, A., G., A.; WIECHETECK, G., K.; BUSCH, O., M., D.; HINSCHING, M., A., O.; PILATTI, F. **Diagnosis of domestic sewage disposal in the Alagados Fountain Basin, Ponta Grossa/PR.** 2008. Available at < http://www.4eetcg.uepg.br/oral/37_1.pdf> Accessed on: 28 Apr. 2012.

JUNIOR, F., R.; CARVALHO, S., L. **Evaluation of Water Quality in the Gavanhery and Lambary Creeks Watershed in the Municipality of Getulina, SP.**

Available at: http://www.periodicos.rc.biblioteca.unesp.br/index.php/holos/article/viewFile/4726/3754 > Accessed on: 28 Apr. 2012.

LAMPARELLI, M.C. **Degree of trophy in bodies of water in the State of São Paulo: evaluation of monitoring methods.** PhD thesis. Institute of Biosciences, University of Sao Paulo. p. 238. 2004.

LEGENDRE, P.; LEGENDRE, L. **Numerical Ecology**. Amsterdam: Elsevier Science. P. 853. 1998.

MANAHAN, S. E. **Environmental Chemistry**. 8th ed., Boca Raton: CRC Press. p. 783. 2005.

MINISTRY OF THE ENVIRONMENT - MMA. **National Environmental Policy 6938/81.** Available at: < http://www.mma.gov.br/estruturas/sqa_pnla/_arquivos/46_10112008050406.pdf >. Accessed on: May 15, 2012.

MINISTRY OF THE ENVIRONMENT - MMA. **National Water Resources Policy 9433/97.** Available at: < http://www.mma.gov.br/estruturas/161/_publicacao/161_publicacao16032012065259 .pdf >. Accessed on: May 15, 2012.

MORAES, E. C. F.; SZNELWAR, R. B.; FERNiCOLA, N. A. G. G. **Manual de toxicologia analitica**. Sao Paulo: Roca, 1991.

NASELLI-FLORES, L. **Limnological aspects of Sicilian reservoirs: a comparative, ecosystemic approach**. In: TUNDISI, J. G; STRASKRABA, M. (Eds). Theoretical reservoir ecology and its applications. Neatherlands, Buckhuys, p. 283311, 1999.

OLIVEIRA, A. T. R. A. Phytoplankton community in the monitoring of rivers in the state of São Paulo. Dissertation (Master's Degree). Sao Paulo: School of Public Health, USP. 2012.

PEREIRA, R. S. Identification and characterization of pollution sources in water systems. **Revista Eletrônica de Recursos Hidricos**, v. 1, p. 23 - 40. 2004.

PETRY, C. F. **Determination of Trace Elements in Environmental Samples by ICP OES.** Porto Alegre, 2005. Dissertation (Master's Degree in Chemistry) - Postgraduate Program in Chemistry, Federal University of Rio Grande do Sul, Porto Alegre - RS, 2005.

PRADO, R. B.; NOVO, E. M. L. M.; PEREIRA, M. N. Proceedings of the 11th Brazilian Symposium on Remote Sensing. Belo Horizonte - MG, p. 2565 - 2567. 2003

QURESHI, A.; MACLEOD, M.; SCHERINGER, M.; HUNGERBÜHLER, K. Mercury cycling and species mass balances in four North American lakes. Safety and Environmental Technology Group, Institute for Chemical and Bioengineering, ETH Zürich, CH-8093 Zürich. **Environmental Pollution** v.157, p. 452-462. 2009

REFOSCO, J., C. **Ecologia da Paisagem e o Sistema de Informações Geográficas no Estudo da Interferência da Paisagem na Concentraçao de Sólidos Totais no Reservatório da Usina de Barra Bonita - SP.** Sao Carlos, 1996. 129f. Dissertation (Master's Degree in Environmental Engineering Sciences) - Sao Carlos School of Engineering, University of Sao Paulo, Sao Carlos - SP. 1996.

RIBEIRO. G. M., MAIA. C.E., MEDEIROS. J.F. **Use of linear regression to estimate the relationship between electrical conductivity and the ionic composition of irrigation water.** Brazilian Journal of Agricultural and Environmental Engineering. Vol. 9. N°1. p.15-22, 2004.

ROSA, A. H., FRACETO, L. F., CARLOS, V. M. Meio Ambiente e Sustentabilidade. **Chap. 5 - Water resources and hydrological indicators.** Porto Alegre, Bookman p. 103 - 125., 2012.

ROSA, A. H., et al. **Scientific Report Editai mct/cnpq/ct-hidro - N° 044/2006/ Proc. 555539/2006-7.** Environmental diagnosis and evaluation of land use and occupation for the sustainability of an important wetland in the Middle Tietê Basin. UNESP - Sorocaba, SP. p.124. 2009.

ROSA, A.H.; ROCHA, J.C.; CARDOSO, A.A. **Introduction to Environmental Chemistry**. Porto Alegre, Bookman, 2004.

SALOMONS, W.; FORSTNER, U. **Metals in the hydrocycle**. Springer-Verlag. p. 349. 1984.

SAMPAIO, S. C.; SILVESTRO M. G.; FRIGO, E. P.; BORGES, C. M. **Relationship between solids series and electrical conductivity in different wastewaters.** Irriga, Botucatu. v. 12, n. 4, p. 557-562, 2007.

SANTOS, R. F. dos. **Environmental planning: theory and practice.** Sao Paulo - Oficina de Textos, 2004.

SCWARTZ, M. E.; KRULL, I. S. Validation of chromatographic methods. **Pharmaceutical Technology**. v. 2, n.3, p.12-20, 1998.

SHUKLA, O.P.; RAI, U.N.; DUBEY, S. Involvement and interaction of microbial communities in the transformation and stabilization of chromium during the composting of tannery effluent treated biomass of Vallisneria spiralis L. **Ecotoxicology and Bioremediation Group**, National Botanical Research Institute, Rana Pratap Marg, Lucknow 226 001, India, 2008.

SILVA, E. R.; ASSIS, O. B. G. Evaluation of electrochemical technique for removal of organic waste from water by using a laboratory scale unit. **Revista Engenharia Sanitària e Ambiental** v.9, p.193-196. 2004.

SILVA, G. A.; GALEMBECK, T., M., B.; GALEMBECK, O. **Underground and Surface Water Availability in the Pirajibu River Sub-Basin,**

Sorocaba-SP Region. VIII Southeast Geology Symposium. Sao Pedro, SP. 2003.

SOARES, F. B. **Environmental planning of the balneàrio da amizade basin in the municipalities of Alvares Machado and Presidente Prudente - Sao Paulo.** Full academic article. Available at: <

http://bacias.fct.unesp.br/gadis/DOCUMENTOS/GestaoDasAguas/Artigos/Artigo_completo_VII%20Forum%20Ambiental%20da%20Alta%20Paulista-3.pdf > Accessed on: May 23, 2012.

SMITH, W. S; PETRERE Jr. M.; BARRELLA, W. **The fish communities of the Sorocaba River basin, State of Sao Paulo, Brazil in different habitats (Sao Paulo, Brazil).** Available at < http://www.scielo.br/scielo.php?script=sci_arttext&pid=S151969842009000500005&nrm=iso&tlng=en> Accessed on: May 15, 2012.

SMITH, W. S.; SALMAZZI, B.A.; POSSOMATO, H.M.; OLIVEIRA L.C.A.; ALMEIDA, M.A.G; PUPO, R.H.; TAVARES, T.A. The Sorocaba River basin: characterization and main impacts. **Revista Cientifica do IMAPES** v. 3, p.110113. 2005

SOUZA, V.V. Net throughfall quality assessment in a spontaneous forest ecosystem (Mata Atlântica) in Viçosa, MG, Brazil. **Revista Arvore** v.31, p.737-743. 2007.

TUNDISI, J.G.; MATSUMURA-TUNDISI, T.. **Limnology.** Sao Paulo, SP: Oficina de Textos, p. 632. 2008.

U.S. ENVIRONMENT PROTECTION AGENCY. (2009a). **Actions You Can Take To Reduce Lead In Drinking Water.** Available at: <www.epa.gov/ogwdw/lead/lead1.html>. Accessed on May 5, 2013.

ULLRICH, S.M., TANTON, T.W., ABDRASHITOVA, S.A. Mercury in the aquatic environment: a review of factors affecting methylation. Critical Reviews in **Environmental Science and Technology** v. 31, p. 241-293. 2001

USEPA (United States Environmental Protection Agency). **Water Quality Criteria Summary.** Office of Science and Technology, Washington, DC, May, 1991.

VALLS, M. and LORENZO, V. Exploiting the genetic and biochemical capacities of bacteria for remediation of heavy metal pollution. **FEMS Microbiology Reviews,** v. 26, p. 327- 338. 2002.

WETZEL, R. G.; LIKENS, G. E. ***Limnological Analyses***. 20 ed., SpringerVerlag. 391 p., 1991.

YANG, M.; SANUDO-WILHELMY, S.A. Cadmium and manganese distributions in the Hudson River Estuary: *interannual and seasonal variability*. **Earth and Planetary Science Letters** v. 160, p. 403-418. 1998

Printed by Books on Demand GmbH, Norderstedt / Germany